Fields Institute Monographs

VOLUME 29

The Fields Institute for Research in Mathematical Sciences

The Fields Institute is a centre for research in the mathematical sciences, located in Toronto, Canada. The Institutes mission is to advance global mathematical activity in the areas of research, education and innovation. The Fields Institute is supported by the Ontario Ministry of Training, Colleges and Universities, the Natural Sciences and Engineering Research Council of Canada, and seven Principal Sponsoring Universities in Ontario (Carleton, McMaster, Ottawa, Toronto, Waterloo, Western and York), as well as by a growing list of Affiliate Universities in Canada, the U.S. and Europe, and several commercial and industrial partners.

For further volumes:
http://www.springer.com/series/10502

Nizar Touzi

Optimal Stochastic Control, Stochastic Target Problems, and Backward SDE

With Chapter 13 by Agnès Tourin

FIELDS The Fields Institute for Research in the Mathematical Sciences

Springer

Nizar Touzi
Département de Mathématiques Appliquées
École Polytechnique
Palaiseau Cedex
France

ISSN 1069-5273 ISSN 2194-3079 (electronic)
ISBN 978-1-4614-4285-1 ISBN 978-1-4614-4286-8 (eBook)
DOI 10.1007/978-1-4614-4286-8
Springer New York Heidelberg Dordrecht London

Library of Congress Control Number: 2012943958

Mathematics Subject Classification (2010): 03C64, 14P15, 26A12, 26A93, 32C05, 32S65, 34C08, 34M40, 37S75, 58A17

Cover illustration: Drawing of J.C. Fields by Keith Yeomans

Printed on acid-free paper

Springer is part of Springer Science+Business Media (www.springer.com)

I should like to express
all my love to my family:
Christine, our sons Ali and Héni, and our daughter Lilia,
who accompanied me during this visit to Toronto,
all my thanks to them for their patience while I was preparing these notes,
and all my apologies for my absence even when I was physically present...

Contents

Chapter 1
Introduction

These notes have been prepared for the graduate course taught at the Fields Institute, Toronto, during the thematic program on quantitative finance which was held from January to June, 2010.

I would like to thank all the participants to these lectures. It was a pleasure for me to share my experience on this subject with the excellent audience that was offered by this special research semester. In particular, their remarks and comments helped to improve parts of this document and to correct some mistakes.

My special thanks go to Bruno Bouchard, Mete Soner, and Agnès Tourin who accepted to act as guest lecturers within this course. These notes have also benefitted from the discussions with them, and some parts are based on my previous work with Bruno and Mete.

These notes have also benefitted from careful reading by Matheus Grasselli, Pierre Henry-Labordère and Tom Salisbury. I greatly appreciate their help and hope there are not many mistakes left.

I would like to express all my thanks to Matheus Grasselli, Tom Hurd, Tom Salisbury, and Sebastian Jaimungal for the warm hospitality at the Fields Institute and their regular attendance to my lectures.

These lectures present the modern approach to stochastic control problems with a special emphasis on the application in financial mathematics. For pedagogical reason, we restrict the scope of the course to the control of diffusion processes, thus ignoring the presence of jumps.

We first review the main tools from stochastic analysis: Brownian motion and the corresponding stochastic integration theory. This already introduces to the first connection with partial differential equations (PDEs). Indeed, by Itô's formula, a linear PDE pops up as the infinitesimal counterpart of the tower property. Conversely, given a nicely behaved smooth solution, the so-called Feynman–Kac formula provides a stochastic representation in terms of a conditional expectation.

N. Touzi, *Optimal Stochastic Control, Stochastic Target Problems, and Backward SDE*, Fields Institute Monographs 29, DOI 10.1007/978-1-4614-4286-8_1,

We then introduce the class of standard stochastic control problems where one wishes to maximize the expected value of some gain functional. The first main task is to derive an original weak dynamic programming principle which avoids the heavy measurable selection arguments in typical proofs of the dynamic programming principle when no a priori regularity of the value function is known. The infinitesimal counterpart of the dynamic programming principle is now a nonlinear PDE which is called dynamic programming equation or Hamilton–Jacobi–Bellman equation. The hope is that the dynamic programming equation provides a complete characterization of the problem, once complemented with appropriate boundary conditions. However, this requires strong smoothness conditions, which can be seen to be violated in simple examples.

A parallel picture can be drawn for optimal stopping problems and, in fact, for the more general control and stopping problems. In these notes we do not treat such mixed control problems, and we rather analyze separately these two classes of control problems. Here again, we derive the dynamic programming principle and the corresponding dynamic programming equation under strong smoothness conditions. In the present case, the dynamic programming equation takes the form of the obstacle problem in PDEs.

When the dynamic programming equation happens to have an explicit smooth solution, the verification argument allows to verify whether this candidate indeed coincides with the value function of the control problem. The verification argument provides as a by-product an access to the optimal control, i.e., the solution of the problem. But of course, such lucky cases are rare, and one should not count on solving any stochastic control problem by verification.

In the absence of any general a priori regularity of the value function, the next development of the theory is based on viscosity solutions. This beautiful notion was introduced by Crandal and Lions and provides a weak notion of solutions to second-order degenerate elliptic PDEs. We review the main tools from viscosity solutions which are needed in stochastic control. In particular, we provide a difficulty-incremental presentation of the comparison result (i.e., maximum principle) which implies uniqueness.

We next show that the weak dynamic programming equation implies that the value function is a viscosity solution of the corresponding dynamic programming equation in a wide generality. In particular, we do not assume that the controls are bounded. We emphasize that in the present setting, there is no a priori regularity of the value function needed to derive the dynamic programming equation: we only need it to be locally bounded! Given the general uniqueness results, viscosity solutions provide a powerful tool for the characterization of stochastic control and optimal stopping problems.

The remaining part of the lectures focus on the more recent literature on stochastic control, namely stochastic target problems. These problems are motivated by the superhedging problem in financial mathematics. Various extensions have been studied in the literature. We focus on a particular setting where the proofs are simplified while highlighting the main ideas.

The use of viscosity solutions is crucial for the treatment of stochastic target problems. Indeed, deriving any a priori regularity seems to be a very difficult task. Moreover, by writing formally the corresponding dynamic programming equation and guessing an explicit solution (in some lucky case), there is no known direct verification argument as in standard stochastic control problems. Our approach is then based on a dynamic programming principle suited to this class of problems, and called geometric dynamic programming principle, due to a further extension of stochastic target problems to front propagation problems in differential geometry. The geometric programming principle allows to obtain a dynamic programming equation in the sense of viscosity solutions. We provide some examples where the analysis of the dynamic programming equation leads to a complete solution of the problem.

We also present an interesting extension to stochastic target problems with controlled probability of success. A remarkable trick allows to reduce these problems to standard stochastic target problems. By using this methodology, we show how one can solve explicitly the problem of quantile hedging which was previously solved by Föllmer and Leukert [21] by duality methods in the standard linear case in financial mathematics.

A further extension of stochastic target problems consists in involving the quadratic variation of the control process in the controlled state dynamics. These problems are motivated by examples from financial mathematics related to market illiquidity and are called second-order stochastic target problems. We follow the same line of arguments by formulating a suitable geometric dynamic programming principle and deriving the corresponding dynamic programming equation in the sense of viscosity solutions. The main new difficulty here is to deal with the short-time asymptotics of double stochastic integrals.

The final part of the lectures explores a special type of stochastic target problems in the non-Markovian framework. This leads to the theory of backward stochastic differential equations (BSDE) which was introduced by Pardoux and Peng [32]. Here, in contrast to stochastic target problems, we insist on the existence of a solution to the stochastic target problem. We provide the main existence, uniqueness, stability, and comparison results. We also establish the connection with stochastic control problems. We finally show the connection with semilinear PDEs in the Markovian case.

The extension of the theory of BSDEs to the case where the generator is quadratic in the control variable is very important in view of the applications to portfolio optimization problems. However, the existence and uniqueness can not be addressed as simply as in the Lipschitz case. The first existence and uniqueness results were established by Kobylanski [26] by adapting to the non-Markovian framework techniques developed in the PDE literature. Instead of this highly technical argument, we report the beautiful argument recently developed by Tevzadze [39] and provide applications in financial mathematics.

The final chapter is dedicated to numerical methods for nonlinear PDEs. We provide a complete proof of convergence based on the Barles–Souganidis motone scheme method. The latter is a beautiful and simple argument which exploits the stability of viscosity solutions. Stronger results are provided in the semilinear case by using techniques from BSDEs.

Chapter 2
Conditional Expectation and Linear Parabolic PDEs

Throughout this chapter, $(\Omega, \mathcal{F}, \mathbb{F}, P)$ is a filtered probability space with filtration $\mathbb{F} = \{\mathcal{F}_t,\ t \geq 0\}$ satisfying the usual conditions. Let $W = \{W_t,\ t \geq 0\}$ be a Brownian motion valued in $\mathbb{R}^d$, defined on $(\Omega, \mathcal{F}, \mathbb{F}, P)$.

Throughout this chapter, a maturity $T > 0$ will be fixed. By $\mathbb{H}^2$, we denote the collection of all progressively measurable processes ϕ with appropriate (finite) dimension such that $\mathbb{E}[\int_0^T |\phi_t|^2 \mathrm{d}t] < \infty$.

2.1 Stochastic Differential Equations

In this section, we recall the basic tools from stochastic differential equations

$$\mathrm{d}X_t = b_t(X_t)\mathrm{d}t + \sigma_t(X_t)\mathrm{d}W_t,\ t \in [0, T], \tag{2.1}$$

where $T > 0$ is a given maturity date. Here, b and σ are $\mathbb{F} \otimes \mathcal{B}(\mathbb{R}^n)$-progressively measurable functions from $[0, T] \times \Omega \times \mathbb{R}^n$ to $\mathbb{R}^n$ and $\mathcal{M}_{\mathbb{R}}(n, d)$, respectively. In particular, for every fixed $x \in \mathbb{R}^n$, the processes $\{b_t(x), \sigma_t(x), t \in [0, T]\}$ are $\mathbb{F}-$progressively measurable.

Definition 2.1. A strong solution of (2.1) is an $\mathbb{F}-$progressively measurable process X such that $\int_0^T (|b_t(X_t)| + |\sigma_t(X_t)|^2)\mathrm{d}t < \infty$, a.s., and

$$X_t = X_0 + \int_0^t b_s(X_s)\mathrm{d}s + \int_0^t \sigma_s(X_s)\mathrm{d}W_s,\quad t \in [0, T].$$

Let us mention that there is a notion of weak solutions, which relaxes some conditions from the above definition in order to allow for more general stochastic differential equations. Weak solutions, as opposed to strong solutions, are defined on some probabilistic structure (which becomes part of the solution), and not necessarily on $(\Omega, \mathcal{F}, \mathbb{F}, \mathbb{P}, W)$. Thus, for a weak solution we search for a probability structure

N. Touzi, *Optimal Stochastic Control, Stochastic Target Problems, and Backward SDE*, Fields Institute Monographs 29, DOI 10.1007/978-1-4614-4286-8_2,

$(\tilde{\Omega}, \tilde{\mathcal{F}}, \tilde{\mathbb{F}}, \tilde{\mathbb{P}}, \tilde{W})$ and a process $\tilde{X}$ such that the requirement of the above definition holds true. Obviously, any strong solution is a weak solution, but the opposite claim is false.

The main existence and uniqueness result is the following.

Theorem 2.2. *Let $X_0 \in \mathbb{L}^2$ be a r.v. independent of W. Assume that the processes $b_.(0)$ and $\sigma_.(0)$ are in $\mathbb{H}^2$ and that for some $K > 0$,*

$$|b_t(x) - b_t(y)| + |\sigma_t(x) - \sigma_t(y)| \leq K|x - y| \quad \textit{for all } t \in [0,T], \ x, y \in \mathbb{R}^n.$$

Then, for all $T > 0$, there exists a unique strong solution of (2.1) *in $\mathbb{H}^2$. Moreover,*

$$\mathbb{E}\left[\sup_{t\leq T} |X_t|^2\right] \leq C\left(1 + \mathbb{E}|X_0|^2\right) e^{CT}, \tag{2.2}$$

for some constant $C = C(T, K)$ depending on T and K.

Proof. We first establish the existence and uniqueness result, then we prove the estimate (2.2).

Step 1. For a constant $c > 0$, to be fixed later, we introduce the norm

$$\|\phi\|_{\mathbb{H}^2_c} := \mathbb{E}\left[\int_0^T e^{-ct}|\phi_t|^2 dt\right]^{1/2} \quad \text{for every } \phi \in \mathbb{H}^2.$$

Clearly, the norms $\|.\|_{\mathbb{H}^2}$ and $\|.\|_{\mathbb{H}^2_c}$ on the Hilbert space $\mathbb{H}^2$ are equivalent. Consider the map U on $\mathbb{H}^2$ by

$$U(X)_t := X_0 + \int_0^t b_s(X_s)ds + \int_0^t \sigma_s(X_s)dW_s, \ 0 \leq t \leq T.$$

By the Lipschitz property of b and σ in the $x-$variable and the fact that $b_.(0)\sigma_.(0) \in \mathbb{H}^2$, it follows that this map is well defined on $\mathbb{H}^2$. In order to prove existence and uniqueness of a solution for (2.1), we shall prove that $U(X) \in \mathbb{H}^2$ for all $X \in \mathbb{H}^2$ and that U is a contracting mapping with respect to the norm $\|.\|_{\mathbb{H}^2_c}$ for a convenient choice of the constant $c > 0$.

1. We first prove that $U(X) \in \mathbb{H}^2$ for all $X \in \mathbb{H}^2$. To see this, we decompose:

$$\|U(X)\|^2_{\mathbb{H}^2} \leq 3T\|X_0\|^2_{\mathbb{L}^2} + 3T\mathbb{E}\left[\int_0^T \left|\int_0^t b_s(X_s)ds\right|^2 dt\right]$$
$$+3\mathbb{E}\left[\int_0^T \left|\int_0^t \sigma_s(X_s)dW_s\right|^2 dt\right].$$

By the Lipschitz continuity of b and σ in x, uniformly in t, we have $|b_t(x)|^2 \le K(1+|b_t(0)|^2+|x|^2)$ for some constant K. We then estimate the second term by:

$$\mathbb{E}\left[\int_0^T \left|\int_0^t b_s(X_s)\mathrm{d}s\right|^2 \mathrm{d}t\right] \le KT\mathbb{E}\left[\int_0^T (1+|b_t(0)|^2+|X_s|^2)\mathrm{d}s\right] < \infty,$$

since $X \in \mathbb{H}^2$, and $b(.,0) \in \mathbb{L}^2([0,T])$.

As, for the third term, we use the Doob maximal inequality together with the fact that $|\sigma_t(x)|^2 \le K(1+|\sigma_t(0)|^2+|x|^2)$, a consequence of the Lipschitz property on σ:

$$\begin{aligned}\mathbb{E}\left[\int_0^T \left|\int_0^t \sigma_s(X_s)\mathrm{d}W_s\right|^2 \mathrm{d}t\right] &\le T\mathbb{E}\left[\max_{t\le T}\left|\int_0^t \sigma_s(X_s)\mathrm{d}W_s\right|^2 \mathrm{d}t\right]\\ &\le 4T\mathbb{E}\left[\int_0^T |\sigma_s(X_s)|^2\mathrm{d}s\right]\\ &\le 4TK\mathbb{E}\left[\int_0^T (1+|\sigma_s(0)|^2+|X_s|^2)\mathrm{d}s\right] < \infty.\end{aligned}$$

2. To see that U is a contracting mapping for the norm $\|.\|_{\mathbb{H}^2_c}$, for some convenient choice of $c>0$, we consider two process $X, Y \in \mathbb{H}^2$ with $X_0 = Y_0$, and we estimate that:

$$\begin{aligned}&\mathbb{E}\,|U(X)_t - U(Y)_t|^2\\ &\le 2\mathbb{E}\left|\int_0^t (b_s(X_s)-b_s(Y_s))\,\mathrm{d}s\right|^2 + 2\mathbb{E}\left|\int_0^t (\sigma_s(X_s)-\sigma_s(Y_s))\,\mathrm{d}W_s\right|^2\\ &= 2\mathbb{E}\left|\int_0^t (b_s(X_s)-b_s(Y_s))\,\mathrm{d}s\right|^2 + 2\mathbb{E}\int_0^t |\sigma_s(X_s)-\sigma_s(Y_s)|^2\,\mathrm{d}s\\ &\le 2t\mathbb{E}\int_0^t |b_s(X_s)-b_s(Y_s)|^2\,\mathrm{d}s + 2\mathbb{E}\int_0^t |\sigma_s(X_s)-\sigma_s(Y_s)|^2\,\mathrm{d}s\\ &\le 2(T+1)K\int_0^t \mathbb{E}\,|X_s-Y_s|^2\,\mathrm{d}s.\end{aligned}$$

Hence, $\|U(X)-U(Y)\|_c \le \dfrac{2K(T+1)}{c}\|X-Y\|_c$, and therefore U is a contracting mapping for sufficiently large c.

Step 2. We next prove the estimate (2.2). We shall alleviate the notation writing $b_s := b_s(X_s)$ and $\sigma_s := \sigma_s(X_s)$. We directly estimate

$$\begin{aligned}\mathbb{E}\left[\sup_{u\le t}|X_u|^2\right] &= \mathbb{E}\left[\sup_{u\le t}\left|X_0+\int_0^u b_s\mathrm{d}s+\int_0^u \sigma_s\mathrm{d}W_s\right|^2\right]\\ &\le 3\left(\mathbb{E}|X_0|^2+t\mathbb{E}\left[\int_0^t |b_s|^2\mathrm{d}s\right]+\mathbb{E}\left[\sup_{u\le t}\left|\int_0^u \sigma_s\mathrm{d}W_s\right|^2\right]\right)\\ &\le 3\left(\mathbb{E}|X_0|^2+t\mathbb{E}\left[\int_0^t |b_s|^2\mathrm{d}s\right]+4\mathbb{E}\left[\int_0^t |\sigma_s|^2\mathrm{d}s\right]\right)\end{aligned}$$

where we used the Doob's maximal inequality. Since b and σ are Lipschitz continuous in x, uniformly in t and ω, this provides:

$$\mathbb{E}\left[\sup_{u\le t}|X_u|^2\right]\le C(K,T)\left(1+\mathbb{E}|X_0|^2+\int_0^t \mathbb{E}\left[\sup_{u\le s}|X_u|^2\right]\mathrm{d}s\right),$$

and we conclude by using the Gronwall lemma. □

The following exercise shows that the Lipschitz-continuity condition on the coefficients b and σ can be relaxed. We observe that further relaxation of this assumption is possible in the one-dimensional case, see, e.g., Karatzas and Shreve [23].

Exercise 2.3. In the context of this section, assume that the coefficients μ and σ are locally Lipschitz and linearly growing in x, uniformly in (t,ω). By a localization argument, prove that strong existence and uniqueness holds for the stochastic differential equation (2.1).

In addition to the estimate (2.2) of Theorem 2.2, we have the following flow continuity results of the solution of the SDE. In order to emphasize the dependence on the initial date, we denote by $\{X_s^{t,x}, s\ge t\}$ the solution of the SDE (2.1) with initial condition $X_t^{t,x}=x$.

Theorem 2.4. *Let the conditions of Theorem 2.2 hold true, and consider some $(t,x)\in[0,T)\times\mathbb{R}^n$ with $t\le t'\le T$.*

(i) *There is a constant C such that*

$$\mathbb{E}\left[\sup_{t\le s\le t'}\left|X_s^{t,x}-X_s^{t,x'}|^2\right|\right]\le C\mathrm{e}^{Ct'}|x-x'|^2. \tag{2.3}$$

(ii) *Assume further that $B:=\sup_{t<t'\le T}(t'-t)^{-1}\mathbb{E}\int_t^{t'}\left(|b_r(0)|^2+|\sigma_r(0)|^2\right)\mathrm{d}r<\infty$. Then for all $t'\in[t,T]$*

$$\mathbb{E}\left[\sup_{t'\le s\le T}\left|X_s^{t,x}-X_s^{t',x}|^2\right|\right]\le C\mathrm{e}^{CT}(B+|x|^2)|t'-t|. \tag{2.4}$$

Proof. (i) To simplify the notations, we set $X_s := X_s^{t,x}$ and $X'_s := X_s^{t,x'}$ for all $s \in [t,T]$. We also denote $\delta x := x - x'$, $\delta X := X - X'$, $\delta b := b(X) - b(X')$ and $\delta\sigma := \sigma(X) - \sigma(X')$. We first decompose

$$|\delta X_s|^2 \le 3\left(|\delta x|^2 + \left|\int_t^s \delta b_u \mathrm{d}u\right|^2 + \left|\int_t^s \delta\sigma_u \mathrm{d}W_u\right|^2\right)$$

$$\le 3\left(|\delta x|^2 + (s-t)\int_t^s \left|\delta b_u\right|^2 \mathrm{d}u + \int_t^s \delta\sigma_u \mathrm{d}W_u\Big|^2\right).$$

Then, it follows from the Doob maximal inequality and the Lipschitz property of the coefficients b and σ that

$$h(t') := \mathbb{E}\left[\sup_{t\le s\le t'} |\delta X_s|^2\right] \le 3\left(|\delta x|^2 + (s-t)\int_t^s \mathbb{E}\left|\delta b_u\right|^2 du\right.$$

$$\left. +4\int_t^s \mathbb{E}\left|\delta\sigma_u\right|^2 d_u\right)$$

$$\le 3\left(|\delta x|^2 + K^2(t'+4)\int_t^s \mathbb{E}|\delta X_u|^2 \mathrm{d}u\right)$$

$$\le 3\left(|\delta x|^2 + K^2(t'+4)\int_t^s h(u)\mathrm{d}u\right).$$

Then the required estimate follows from the Gronwall inequality.

(ii) We next prove (2.4). We again simplify the notation by setting $X_s := X_s^{t,x}$, $s \in [t,T]$, and $X'_s := X_s^{t',x}$, $s \in [t',T]$. We also denote $\delta t := t' - t$, $\delta X := X - X'$, $\delta b := b(X) - b(X')$, and $\delta\sigma := \sigma(X) - \sigma(X')$. Then following the same arguments as in the previous step, we obtain for all $u \in [t',T]$:

$$h(u) := \mathbb{E}\left[\sup_{t'\le s\le u} |\delta X_s|^2\right] \le 3\left(\mathbb{E}|X_{t'} - x|^2 + K^2(T+4)\int_{t'}^u \mathbb{E}|\delta X_r|^2 \mathrm{d}r\right)$$

$$\le 3\left(\mathbb{E}|X_{t'} - x|^2 + K^2(T+4)\int_{t'}^u h(r)\mathrm{d}r\right) \tag{2.5}$$

Observe that

$$\mathbb{E}|X_{t'} - x|^2 \le 2\left(\mathbb{E}\left|\int_t^{t'} b_r(X_r)\mathrm{d}r\right|^2 + \mathbb{E}\left|\int_t^{t'} \sigma_r(X_r)\mathrm{d}r\right|^2\right)$$

$$\le 2\left(T\int_t^{t'} \mathbb{E}|b_r(X_r)|^2 \mathrm{d}r + \int_t^{t'} \mathbb{E}|\sigma_r(X_r)|^2 \mathrm{d}r\right)$$

$$\leq 6(T+1)\int_t^{t'} \left(K^2\mathbb{E}|X_r - x|^2 + |x|^2 + \mathbb{E}|b_r(0)|^2\right)\mathrm{d}r$$

$$\leq 6(T+1)\Big((t'-t)(|x|^2+B) + K^2\int_t^{t'} \mathbb{E}|X_r - x|^2\mathrm{d}r\Big).$$

By the Gronwall inequality, this shows that

$$\mathbb{E}|X_{t'} - x|^2 \leq C(|x|^2+B)|t'-t|\mathrm{e}^{C(t'-t)}.$$

Plugging this estimate in (2.5), we see that

$$h(u) \leq 3\left(C(|x|^2+B)|t'-t|\mathrm{e}^C(t'-t) + K^2(T+4)\int_{t'}^{u} h(r)\mathrm{d}r\right), \tag{2.6}$$

and the required estimate follows from the Gronwall inequality. □

2.2 Markovian Solutions of SDEs

In this section, we restrict the coefficients b and σ to be deterministic functions of (t,x). In this context, we write

$$b_t(x) = b(t,x),\ \sigma_t(x) = \sigma(t,x) \text{ for } t\in[0,T],\ x\in\mathbb{R}^n,$$

where b and σ are continuous functions, Lipschitz in x uniformly in t. Let $X^{t,x}_{.}$ denote the solution of the stochastic differential equation

$$X^{t,x}_s = x + \int_t^s b\big(u, X^{t,x}_u\big)\mathrm{d}u + \int_t^s \sigma\big(u, X^{t,x}_u\big)\mathrm{d}W_u\ \ s\geq t.$$

The two following properties are obvious:

- Clearly, $X^{t,x}_s = F\left(t,x,s,(W_. - W_t)_{t\leq u\leq s}\right)$ for some deterministic function F.
- For $t\leq u\leq s$: $X^{t,x}_s = X^{u,X^{t,x}_u}_s$. This follows from the pathwise uniqueness, and holds also when u is a stopping time.

With these observations, we have the following Markovian property for the solutions of stochastic differential equations.

Proposition 2.5 (Strong Markovian property). *For all stopping time τ with values in [0, s]*

$$\mathbb{E}\left[\Phi\left(X_u, \tau\leq u\leq s\right)|\mathcal{F}_\tau\right] = \mathbb{E}\left[\Phi\left(X_u, \tau\leq u\leq s\right)|X_\tau\right]$$

for all bounded function $\Phi:\ C([\tau,s])\longrightarrow\mathbb{R}$.

2.3 Connection with Linear Partial Differential Equations

2.3.1 Generator

Let $\{X_s^{t,x}, s \geq t\}$ be the unique strong solution of

$$X_s^{t,x} = x + \int_t^s b(u, X_u^{t,x})\mathrm{d}u + \int_t^s \sigma(u, X_u^{t,x})\mathrm{d}W_u, \ s \geq t,$$

where μ and σ satisfy the required condition for existence and uniqueness of a strong solution.

For a function $f : \mathbb{R}^n \longrightarrow \mathbb{R}$, we define the function $\mathcal{A}f$ by

$$\mathcal{A}f(t,x) = \lim_{h \to 0} \frac{\mathbb{E}[f(X_{t+h}^{t,x})] - f(x)}{h} \quad \text{if the limit exists.}$$

Clearly, $\mathcal{A}f$ is well defined for all bounded C^2- function with bounded derivatives and

$$\mathcal{A}f(t,x) = b(t,x) \cdot Df(x) + \frac{1}{2}\mathrm{Tr}\left[\sigma\sigma^{\mathrm{T}}(t,x)D^2f(x)\right], \tag{2.7}$$

where Df and D^2f denote the gradient and Hessian of f, respectively. (Exercise !). The linear differential operator $\mathcal{A}$ is called the *generator* of X. It turns out that the process X can be completely characterized by its generator or, more precisely, by the generator and the corresponding domain of definition.

As the following result shows, the generator provides an intimate connection between conditional expectations and linear partial differential equations.

Proposition 2.6. *Assume that the function* $(t,x) \longmapsto v(t,x) := \mathbb{E}\left[g(X_T^{t,x})\right]$ *is* $C^{1,2}([0,T) \times \mathbb{R}^n)$. *Then* v *solves the partial differential equation:*

$$\frac{\partial v}{\partial t} + \mathcal{A}v = 0 \ \textit{and} \ v(T,.) = g.$$

Proof. Given (t,x) with $2\delta := T - t > 0$, let $\tau_1 := (T - \delta) \wedge \inf\{u > t : |X_u^{t,x} - x| \geq 1\}$. For an arbitrary $s > t$, it follows from the law of iterated expectation together with the Markovian property of the process X that

$$v(t,x) = \mathbb{E}\left[v\left(s \wedge \tau_1, X_{s\wedge\tau_1}^{t,x}\right)\right].$$

Since $v \in C^{1,2}([0,T), \mathbb{R}^n)$, we may apply Itô's formula, and we obtain by taking expectations:

$$
\begin{aligned}
0 &= \mathbb{E}\left[\int_t^{s\wedge\tau_1}\left(\frac{\partial v}{\partial t}+\mathcal{A}v\right)(u,X_u^{t,x})\mathrm{d}u\right] \\
&\quad +\mathbb{E}\left[\int_t^{s\wedge\tau_1}\frac{\partial v}{\partial x}(u,X_s^{t,x})\cdot\sigma(u,X_u^{t,x})\mathrm{d}W_u\right] \\
&= \mathbb{E}\left[\int_t^{s\wedge\tau_1}\left(\frac{\partial v}{\partial t}+\mathcal{A}v\right)(u,X_u^{t,x})\mathrm{d}u\right],
\end{aligned}
$$

where the last equality follows from the boundedness of $(u, X_u^{t,x})$ on $[t, s \wedge \tau_1]$ together with the continuity of σ. We now send $s \searrow t$, and the required result follows from the dominated convergence theorem. □

2.3.2 *Cauchy Problem and the Feynman–Kac Representation*

In this section, we consider the following linear partial differential equation,

$$
\begin{aligned}
&\frac{\partial v}{\partial t}+\mathcal{A}v-k(t,x)v+f(t,x)=0,\quad (t,x)\in[0,T)\times\mathbb{R}^d \\
&v(T,.)=g
\end{aligned}
, \tag{2.8}
$$

where $\mathcal{A}$ is the generator (2.7), g is a given function from $\mathbb{R}^d$ to $\mathbb{R}$, k and f are functions from $[0,T]\times\mathbb{R}^d$ to $\mathbb{R}$, b and σ are functions from $[0,T]\times\mathbb{R}^d$ to $\mathbb{R}^d$ and $\mathcal{M}_{\mathbb{R}}(d,d)$, respectively. This is the so-called Cauchy problem.

For example, when $k = f \equiv 0$, $b \equiv 0$, and σ is the identity matrix, the above partial differential equation reduces to the heat equation.

Our objective is to provide a representation of this purely deterministic problem by means of stochastic differential equations. We then assume that b and σ satisfy the conditions of Theorem 2.2, namely that

$$
b,\sigma \text{ Lipschitz in } x \text{ uniformly in } t,\quad \int_0^T\left(|b(t,0)|^2+|\sigma(t,0)|^2\right)\mathrm{d}t<\infty. \tag{2.9}
$$

Theorem 2.7. *Let the coefficients $b\sigma$ be continuous and satisfy* (2.9). *Assume further that the function k is uniformly bounded from below and f has quadratic growth in x uniformly in t. Let v be a $C^{1,2}\left([0,T),\mathbb{R}^d\right)$ solution of* (2.8) *with quadratic growth in x uniformly in t. Then*

$$
v(t,x) = \mathbb{E}\left[\int_t^T \beta_s^{t,x} f(s,X_s^{t,x})\mathrm{d}s+\beta_T^{t,x}g\left(X_T^{t,x}\right)\right],\quad t\le T,\ x\in\mathbb{R}^d,
$$

where $\{X_s^{t,x}, s\ge t\}$ is the solution of the SDE (2.1) *with initial data $X_t^{t,x}=x$, and $\beta_s^{t,x} := \mathrm{e}^{-\int_t^s k(u,X_u^{t,x})\mathrm{d}u}$ for $t\le s\le T$.*

Proof. For $n \geq 1$, we denote $T_n := T - 1/n$. For $t < T$, we have $T_n > t$ for large n, and we introduce the sequence of stopping times

$$\tau_n := T_n \wedge \inf\{s > t \ : \ |X_s^{t,x} - x| \geq n\}.$$

Then $\tau_n \longrightarrow T$ $\mathbb{P}$–a.s. Since v is smooth, it follows from Itô's formula that for $t \leq s < T$

$$\begin{aligned} d\left(\beta_s^{t,x} v\left(s, X_s^{t,x}\right)\right) &= \beta_s^{t,x}\left(-kv + \frac{\partial v}{\partial t} + \mathcal{A}v\right)\left(s, X_s^{t,x}\right)\mathrm{d}s \\ &\quad + \beta_s^{t,x}\frac{\partial v}{\partial x}\left(s, X_s^{t,x}\right)\cdot\sigma\left(s, X_s^{t,x}\right)\mathrm{d}W_s \\ &= \beta_s^{t,x}\left(-f(s, X_s^{t,x})\mathrm{d}s + \frac{\partial v}{\partial x}\left(s, X_s^{t,x}\right)\cdot\sigma\left(s, X_s^{t,x}\right)\mathrm{d}W_s\right), \end{aligned}$$

by the PDE satisfied by v in (2.8). Then:

$$\begin{aligned} &\mathbb{E}\left[\beta_{\tau_n}^{t,x} v\left(\tau_n, X_{\tau_n}^{t,x}\right)\right] - v(t,x) \\ &\quad = \mathbb{E}\left[\int_t^{\tau_n} \beta_s^{t,x}\left(-f(s, X_s)\mathrm{d}s + \frac{\partial v}{\partial x}\left(s, X_s^{t,x}\right)\cdot\sigma\left(s, X_s^{t,x}\right)\mathrm{d}W_s\right)\right]. \end{aligned}$$

Now observe that the integrands in the stochastic integral are bounded by definition of the stopping time τ_n, the smoothness of v, and the continuity of σ and the boundedness of β. Then the stochastic integral has zero mean, and we deduce that

$$v(t,x) = \mathbb{E}\left[\int_t^{\tau_n} \beta_s^{t,x} f\left(s, X_s^{t,x}\right)\mathrm{d}s + \beta_{\tau_n}^{t,x} v\left(\tau_n, X_{\tau_n}^{t,x}\right)\right]. \tag{2.10}$$

Since $\tau_n \longrightarrow T$ and the Brownian motion has continuous sample paths $\mathbb{P}$–a.s., it follows from the continuity of v that $\mathbb{P}$–a.s.

$$\begin{aligned} &\int_t^{\tau_n} \beta_s^{t,x} f\left(s, X_s^{t,x}\right)\mathrm{d}s + \beta_{\tau_n}^{t,x} v\left(\tau_n, X_{\tau_n}^{t,x}\right) \\ &\quad \overset{n\to\infty}{\longrightarrow} \int_t^T \beta_s^{t,x} f\left(s, X_s^{t,x}\right)\mathrm{d}s + \beta_T^{t,x} v\left(T, X_T^{t,x}\right) \\ &\quad = \int_t^T \beta_s^{t,x} f\left(s, X_s^{t,x}\right)\mathrm{d}s + \beta_T^{t,x} g\left(X_T^{t,x}\right) \end{aligned} \tag{2.11}$$

by the terminal condition satisfied by v in (2.8). Moreover, since k is bounded from below and the functions f and v have quadratic growth in x uniformly in t, we have

$$\left|\int_t^{\tau_n} \beta_s^{t,x} f\left(s, X_s^{t,x}\right)\mathrm{d}s + \beta_{\tau_n}^{t,x} v\left(\tau_n, X_{\tau_n}^{t,x}\right)\right| \leq C\left(1 + \max_{t \leq T}|X_t|^2\right).$$

By the estimate stated in the existence and uniqueness Theorem 2.2, the latter bound is integrable, and we deduce from the dominated convergence theorem that the convergence in (2.11) holds in $\mathbb{L}^1(\mathbb{P})$, proving the required result by taking limits in (2.10). □

The above Feynman–Kac representation formula has an important numerical implication. Indeed it opens the door to the use of Monte Carlo methods in order to obtain a numerical approximation of the solution of the partial differential equation (2.8). For the sake of simplicity, we provide the main idea in the case $f = k = 0$. Let $\left(X^{(1)}, \ldots, X^{(k)}\right)$ be an iid sample drawn in the distribution of $X_T^{t,x}$, and compute the mean:

$$\hat{v}_k(t,x) := \frac{1}{k}\sum_{i=1}^{k} g\left(X^{(i)}\right).$$

By the law of large numbers, it follows that $\hat{v}_k(t,x) \longrightarrow v(t,x)$ $\mathbb{P}$–a.s. Moreover, the error estimate is provided by the central limit theorem:

$$\sqrt{k}\left(\hat{v}_k(t,x) - v(t,x)\right) \overset{k\to\infty}{\longrightarrow} \mathcal{N}\left(0, \mathbb{V}ar\left[g\left(X_T^{t,x}\right)\right]\right) \text{ in distribution}$$

and is remarkably independent of the dimension d of the variable X!

2.3.3 Representation of the Dirichlet Problem

Let D be an open subset of $\mathbb{R}^d$. The *Dirichlet problem* is to find a function u solving:

$$\mathcal{A}u - ku + f = 0 \text{ on } D \text{ and } u = g \text{ on } \partial D, \tag{2.12}$$

where ∂D denotes the boundary of D and $\mathcal{A}$ is the generator of the process $X^{0,x}$ defined as the unique strong solution of the stochastic differential equation

$$X_t^{0,x} = x + \int_0^t \mu(X_s^{0,x})\mathrm{d}s + \int_0^t \sigma(X_s^{0,x})\mathrm{d}W_s,\ t \geq 0.$$

Similarly to the representation result of the Cauchy problem obtained in Theorem 2.7, we have the following representation result for the Dirichlet problem.

Theorem 2.8. *Let u be a C^2–solution of the Dirichlet problem* (2.12). *Assume that k is nonnegative, and*

$$\mathbb{E}[\tau_D^x] < \infty,\ x \in \mathbb{R}^d,\ \textit{where } \tau_D^x := \inf\left\{t \geq 0 :\ X_t^{0,x} \notin D\right\}.$$

Then, we have the representation

$$u(x) = \mathbb{E}\left[g\left(X^{0,x}_{\tau^x_D}\right)e^{-\int_0^{\tau^x_D} k(X_s)\mathrm{d}s} + \int_0^{\tau^x_D} f\left(X^{0,x}_t\right)e^{-\int_0^t k(X_s)\mathrm{d}s}\mathrm{d}t \right].$$

Exercise 2.9. Provide a proof of Theorem 2.8 by imitating the arguments in the proof of Theorem 2.7.

2.4 The Black–Scholes Model

In this section, we provide a self-contained review of the standard theory of pricing and hedging derivative securities in complete markets. In the context of a Markovian model for the underlying assets, the solution is expressed in terms of a parabolic PDE or, equivalently, a conditional expectation operator.

2.4.1 The Continuous-Time Financial Market

Let T be a finite horizon and $(\Omega, \mathcal{F}, \mathbb{P})$ be a complete probability space supporting a Brownian motion $W = \{(W_t^1, \ldots, W_t^d), 0 \le t \le T\}$ with values in $\mathbb{R}^d$. We denote by $\mathbb{F} = \mathbb{F}^W = \{\mathcal{F}_t, 0 \le t \le T\}$ the canonical augmented filtration of W, i.e., the canonical filtration augmented by zero measure sets of $\mathcal{F}_T$.

We consider a financial market consisting of $d + 1$ assets:

(i) The first asset S^0 is locally riskless, and is defined by

$$S_t^0 = \exp\left(\int_0^t r_u \mathrm{d}u\right), \quad 0 \le t \le T,$$

where $\{r_t, t \in [0,T]\}$ is a non-negative adapted processes with $\int_0^T r_t \mathrm{d}t < \infty$ a.s. and represents the instantaneous interest rate.

(ii) The d remaining assets S^i, $i = 1, \ldots, d$, are risky assets with price processes defined by the dynamics

$$\frac{\mathrm{d}S_t^i}{S_t^i} = \mu_t^i \mathrm{d}t + \sum_{j=1}^d \sigma_t^{i,j} \mathrm{d}W_t^j, \ t \in [0,T],$$

for $1 \le i \le d$, where μ, σ are $\mathbb{F}$–adapted processes with $\int_0^T |\mu_t^i| \mathrm{d}t + \int_0^T |\sigma_t^{i,j}|^2 \mathrm{d}t < \infty$ for all $i, j = 1, \ldots, d$. It is convenient to use the matrix notations to represent the dynamics of the price vector $S = (S^1, \ldots, S^d)$:

$$dS_t = S_t \star (\mu_t dt + \sigma_t dW_t), \ t \in [0, T],$$

where, for two vectors $x, y \in \mathbb{R}^d$, we denote $x \star y$ the vector of $\mathbb{R}^d$ with components $(x \star y)_i = x_i y_i$, $i = 1, \ldots, d$, and μ, σ are the $\mathbb{R}^d$–vector with components μ^i's, and the $\mathcal{M}_{\mathbb{R}}(d, d)$–matrix with entries $\sigma^{i,j}$.

We assume that the $\mathcal{M}_{\mathbb{R}}(d, d)$–matrix σ_t is invertible for every $t \in [0, T]$ a.s., and we introduce the process

$$\lambda_t := \sigma_t^{-1}(\mu_t - r_t \mathbf{1}), \ 0 \le t \le T,$$

called the *risk premium process*. Here $\mathbf{1}$ is the vector of ones in $\mathbb{R}^d$. We shall frequently make use of the discounted processes

$$\tilde{S}_t := \frac{S_t}{S_t^0} = S_t \exp\left(-\int_0^t r_u du\right)$$

Using the above matrix notations, the dynamics of the process $\tilde{S}$ is given by

$$d\tilde{S}_t = \tilde{S}_t \star \big((\mu_t - r_t \mathbf{1})dt + \sigma_t dW_t\big) = \tilde{S}_t \star \sigma_t (\lambda_t dt + dW_t).$$

2.4.2 *Portfolio and Wealth Process*

A portfolio strategy is an $\mathbb{F}$–adapted process $\pi = \{\pi_t, 0 \le t \le T\}$ with values in $\mathbb{R}^d$. For $1 \le i \le n$ and $0 \le t \le T$, π_t^i is the amount (in Euros) invested in the risky asset S^i.

We next recall the self-financing condition in the present framework. Let X_t^π denote the portfolio value, or wealth, process at time t induced by the portfolio strategy π. Then, the amount invested in the non-risky asset is $X_t^\pi - \sum_{i=1}^n \pi_t^i = X_t^\pi - \pi_t \cdot \mathbf{1}$, and the dynamics of the wealth process is given by

$$dX_t^\pi = \sum_{i=1}^n \frac{\pi_t^i}{S_t^i} dS_t^i + \frac{X_t^\pi - \pi_t \cdot \mathbf{1}}{S_t^0} dS_t^0.$$

Let $\tilde{X}^\pi$ be the discounted wealth process

$$\tilde{X}_t^\pi := X_t^\pi \exp\left(-\int_0^t r_u du\right), \ 0 \le t \le T.$$

Then, by an immediate application of Itô's formula, we see that

$$d\tilde{X}_t = \tilde{\pi}_t \cdot \sigma_t (\lambda_t dt + dW_t), \ 0 \le t \le T, \tag{2.13}$$

where $\tilde{\pi}_t := \mathrm{e}^{-\int_0^t r_u \mathrm{d}u} \pi_t$. We still need to place further technical conditions on π, at least in order for the above wealth process to be well defined as a stochastic integral.

Before this, let us observe that, assuming that the risk premium process satisfies the Novikov condition:

$$\mathbb{E}\left[\mathrm{e}^{\frac{1}{2}\int_0^T |\lambda_t|^2 \mathrm{d}t}\right] < \infty,$$

it follows from the Girsanov theorem that the process

$$B_t := W_t + \int_0^t \lambda_u \mathrm{d}u, \quad 0 \le t \le T, \tag{2.14}$$

is a Brownian motion under the equivalent probability measure

$$\mathbb{Q} := Z_T \cdot \mathbb{P} \text{ on } \mathcal{F}_T \text{ where } Z_T := \exp\left(-\int_0^T \lambda_u \cdot \mathrm{d}W_u - \frac{1}{2}\int_0^T |\lambda_u|^2 \mathrm{d}u\right).$$

In terms of the $\mathbb{Q}$ Brownian motion B, the discounted price process satisfies

$$\mathrm{d}\tilde{S}_t = \tilde{S}_t \star \sigma_t \mathrm{d}B_t, \quad t \in [0, T],$$

and the discounted wealth process induced by an initial capital X_0 and a portfolio strategy π can be written in

$$\tilde{X}_t^\pi = \tilde{X}_0 + \int_0^t \tilde{\pi}_u \cdot \sigma_u \mathrm{d}B_u, \quad \text{for } 0 \le t \le T. \tag{2.15}$$

Definition 2.10. An admissible portfolio process $\pi = \{\pi_t, t \in [0, T]\}$ is an $\mathbb{F}$–progressively measurable process such that $\int_0^T |\sigma_t^{\mathrm{T}} \pi_t|^2 \mathrm{d}t < \infty$, a.s., and the corresponding discounted wealth process is bounded from below by a $\mathbb{Q}$–martingale

$$\tilde{X}_t^\pi \ge M_t^\pi, \; 0 \le t \le T, \quad \text{for some } \mathbb{Q}\text{–martingale } M^\pi.$$

The collection of all admissible portfolio processes will be denoted by $\mathcal{A}$.

The lower bound M^π, which may depend on the portfolio π, has the interpretation of a finite credit line imposed on the investor. This natural generalization of the more usual constant credit line corresponds to the situation where the total credit available to an investor is indexed by some financial holding, such as the physical assets of the company or the personal home of the investor, used as collateral. From the mathematical viewpoint, this condition is needed in order to exclude any arbitrage opportunity and will be justified in the subsequent subsection.

2.4.3 Admissible Portfolios and No-Arbitrage

We first define precisely the notion of no-arbitrage.

Definition 2.11. We say that the financial market contains no-arbitrage opportunities if for any admissible portfolio process $\theta \in \mathcal{A}$,

$$X_0 = 0 \text{ and } X_T^\theta \geq 0 \; \mathbb{P} - \text{a.s. implies } X_T^\theta \;=\; 0 \; \mathbb{P} - \text{a.s.}$$

The purpose of this section is to show that the financial market described above contains no-arbitrage opportunities. Our first observation is that, by the very definition of the probability measure $\mathbb{Q}$, the discounted price process $\tilde{S}$ satisfies:

$$\text{the process } \left\{\tilde{S}_t, \; 0 \leq t \leq T\right\} \text{ is a } \mathbb{Q} - \text{local martingale.} \tag{2.16}$$

For this reason, $\mathbb{Q}$ is called a *risk neutral measure*, or an *equivalent local martingale measure*, for the price process S.

We also observe that the discounted wealth process satisfies:

$$\tilde{X}^\pi \text{ is a } \mathbb{Q}\text{–local martingale for every } \pi \in \mathcal{A}, \tag{2.17}$$

as a stochastic integral with respect to the $\mathbb{Q}$–Brownian motion B.

Theorem 2.12. *The continuous-time financial market described above contains no-arbitrage opportunities, i.e., for every $\pi \in \mathcal{A}$*

$$X_0 = 0 \text{ and } X_T^\pi \geq 0 \; \mathbb{P} - a.s. \Longrightarrow X_T^\pi = 0 \; \mathbb{P} - a.s.$$

Proof. For $\pi \in \mathcal{A}$, the discounted wealth process $\tilde{X}^\pi$ is a $\mathbb{Q}$–local martingale bounded from below by a $\mathbb{Q}$–martingale. Then $\tilde{X}^\pi$ is a $\mathbb{Q}$–super-martingale (Exercise!). In particular, $\mathbb{E}^{\mathbb{Q}}\left[\tilde{X}_T^\pi\right] \leq \tilde{X}_0 = X_0$. Recall that $\mathbb{Q}$ is equivalent to $\mathbb{P}$ and S^0 is strictly positive. Then, this inequality shows that, whenever $X_0^\pi = 0$ and $X_T^\pi \geq 0$ $\mathbb{P}$–a.s. (or equivalently $\mathbb{Q}$–a.s.), we have $\tilde{X}_T^\pi = 0$ $\mathbb{Q}$–a.s. and therefore $X_T^\pi = 0$ $\mathbb{P}$–a.s. □

2.4.4 Super-Hedging and No-Arbitrage Bounds

Let G be an $\mathcal{F}_T$–measurable random variable representing the payoff of a derivative security with given maturity $T > 0$. The *super hedging* problem consists in finding the minimal initial cost so as to be able to face the payment G without risk at the maturity T of the contract:

$$V(G) := \inf\{X_0 \in \mathbb{R} \; : \; X_T^\pi \geq G \; \mathbb{P} - \text{a.s. for some } \pi \in \mathcal{A}\}\,.$$

Remark 2.13. Notice that $V(G)$ depends on the reference measure $\mathbb{P}$ only by means of the corresponding null sets. Therefore, the super-hedging problem is not changed if $\mathbb{P}$ is replaced by any equivalent probability measure.

We now show that, under the no-arbitrage condition, the super hedging problem provides *no-arbitrage bounds* on the market price of the derivative security.

Assume that the buyer of the contingent claim G has the same access to the financial market than the seller. Then $V(G)$ *is the maximal amount that the buyer of the contingent claim contract is willing to pay*. Indeed, if the seller requires a premium of $V(G) + 2\varepsilon$, for some $\varepsilon > 0$, then the buyer would not accept to pay this amount as he can obtain at least G by trading on the financial market with initial capital $V(G) + \varepsilon$.

Now, since selling of the contingent claim G is the same as buying the contingent claim $-G$, we deduce from the previous argument that

$$-V(-G) \leq \text{ market price of } G \leq V(G). \tag{2.18}$$

2.4.5 The No-Arbitrage Valuation Formula

We denote by $p(G)$ the market price of a derivative security G.

Theorem 2.14. *Let G be an $\mathcal{F}_T$–measurable random variable representing the payoff of a derivative security at the maturity $T > 0$, and recall the notation $\tilde{G} := G\exp\left(-\int_0^T r_t \mathrm{d}t\right)$. Assume that $\mathbb{E}^{\mathbb{Q}}[|\tilde{G}|] < \infty$. Then, in the context of our (complete) financial market:*

$$p(G) = V(G) = \mathbb{E}^{\mathbb{Q}}[\tilde{G}].$$

Moreover, there exists a portfolio $\pi^* \in \mathcal{A}$ such that $X_0^{\pi^*} = p(G)$ and $X_T^{\pi^*} = G$, a.s., i.e., π^* is a perfect replication strategy.

Proof. 1. We first prove that $V(G) \geq \mathbb{E}^{\mathbb{Q}}[\tilde{G}]$. Let X_0 and $\pi \in \mathcal{A}$ be such that $X_T^{\pi} \geq G$, a.s. or, equivalently, $\tilde{X}_T^{\pi} \geq \tilde{G}$ a.s. Notice that $\tilde{X}^{\pi}$ is a $\mathbb{Q}$–super-martingale, as a $\mathbb{Q}$–local martingale bounded from below by a $\mathbb{Q}$–martingale. Then $X_0 = \tilde{X}_0 \geq \mathbb{E}^{\mathbb{Q}}[\tilde{X}_T^{\pi}] \geq \mathbb{E}^{\mathbb{Q}}[\tilde{G}]$.

2. We next prove that $V(G) \leq \mathbb{E}^{\mathbb{Q}}[\tilde{G}]$. Define the $\mathbb{Q}$–martingale $Y_t := \mathbb{E}^{\mathbb{Q}}[\tilde{G}|\mathcal{F}_t]$ and observe that $\mathbb{F}^W = \mathbb{F}^B$. Then, it follows from the martingale representation theorem that $Y_t = Y_0 + \int_0^T \phi_t \cdot \mathrm{d}B_t$ for some $\mathbb{F}$–adapted process ϕ with $\int_0^T |\phi_t|^2 \mathrm{d}t < \infty$ a.s. Setting $\tilde{\pi}^* := (\sigma^{\mathrm{T}})^{-1}\phi$, we see that

$$\pi^* \in \mathcal{A} \text{ and } Y_0 + \int_0^T \tilde{\pi}^* \cdot \sigma_t \mathrm{d}B_t = \tilde{G} \ \mathbb{P} - \text{a.s.}$$

which implies that $Y_0 \geq V(G)$ and π^* is a perfect hedging strategy for G, starting from the initial capital Y_0.

3. From the previous steps, we have $V(G) = \mathbb{E}^{\mathbb{Q}}[\tilde{G}]$. Applying this result to $-G$, we see that $V(-G) = -V(G)$, so that the no-arbitrage bounds (2.18) imply that the no-arbitrage market price of G is given by $V(G)$. □

2.4.6 PDE Characterization of the Black–Scholes Price

In this subsection, we specialize further the model to the case where the risky securities price processes are Markovian diffusions defined by the stochastic differential equations:

$$\mathrm{d}S_t = S_t \star \big(r(t, S_t)\mathrm{d}t + \sigma(t, S_t)\mathrm{d}B_t\big).$$

Here $(t, s) \longmapsto s \star r(t, s)$ and $(t, s) \longmapsto s \star \sigma(t, s)$ are functions from $\mathbb{R}_+ \times [0, \infty)^d$ to $\mathbb{R}^d$ and $\mathcal{S}_d$, successively satisfying the required continuity and Lipschitz conditions. We also consider a *Vanilla* derivative security defined by the payoff

$$G = g(S_T),$$

where $g : [0, \infty)^d \to \mathbb{R}$ is a measurable function with linear growth. By an immediate extension of the results from the previous subsection, the no-arbitrage price at time t of this derivative security is given by

$$V(t, S_t) = \mathbb{E}^{\mathbb{Q}}\Big[\mathrm{e}^{-\int_t^T r(u,S_u)\mathrm{d}u} g(S_T)|\mathcal{F}_t\Big] = \mathbb{E}^{\mathbb{Q}}\Big[\mathrm{e}^{-\int_t^T r(u,S_u)\mathrm{d}u} g(S_T)|S_t\Big],$$

where the last equality follows from the Markovian property of the process S. Since g has linear growth, it follows that V has linear growth in s uniformly in t. Since V is defined by a conditional expectation, it is expected to satisfy the linear PDE:

$$-\partial_t V - rs \star DV - \frac{1}{2}\mathrm{Tr}\big[(s \star \sigma)^2 D^2 V\big] + rV = 0. \tag{2.19}$$

More precisely, if $V \in C^{1,2}(\mathbb{R}_+, \mathbb{R}^d)$, then V is a classical solution of (2.19) and satisfies the final condition $V(T, .) = g$. Conversely, if the PDE (2.19) combined with the final condition $v(T, .) = g$ has a classical solution v with linear growth, then v coincides with the derivative security price V as stated by Theorem 2.7.

Chapter 3
Stochastic Control and Dynamic Programming

In this chapter, we assume that the filtration $\mathbb{F}$ is the $\mathbb{P}-$augmentation of the canonical filtration of the Brownian motion W. This restriction is only needed in order to simplify the presentation of the proof of the dynamic programming principle. We will also denote by

$$\mathbf{S} := [0,T) \times \mathbb{R}^d \text{ where } T \in [0,\infty].$$

The set $\mathbf{S}$ is called the *parabolic* interior of the state space. We will denote by $\bar{\mathbf{S}} := \mathrm{cl}(\mathbf{S})$ its closure, i.e., $\bar{\mathbf{S}} = [0,T] \times \mathbb{R}^d$ for finite T and $\bar{\mathbf{S}} = \mathbf{S}$ for $T = \infty$.

3.1 Stochastic Control Problems in Standard Form

Control Processes. Given a subset U of $\mathbb{R}^k$, we denote by $\mathcal{U}$ the set of all progressively measurable processes $\nu = \{\nu_t,\ t < T\}$ valued in U. The elements of $\mathcal{U}$ are called control processes.

Controlled process. Let

$$b\ :\ (t,x,u) \in \mathbf{S} \times U \longrightarrow b(t,x,u) \in \mathbb{R}^d$$

and

$$\sigma\ :\ (t,x,u) \in \mathbf{S} \times U \longrightarrow \sigma(t,x,u) \in \mathcal{M}_{\mathbb{R}}(n,d)$$

be two continuous functions satisfying the conditions

$$|b(t,x,u) - b(t,y,u)| + |\sigma(t,x,u) - \sigma(t,y,u)| \le K\ |x-y|, \tag{3.1}$$

$$|b(t,x,u)| + |\sigma(t,x,u)| \le K\ (1 + |x| + |u|) \tag{3.2}$$

N. Touzi, *Optimal Stochastic Control, Stochastic Target Problems, and Backward SDE*,
Fields Institute Monographs 29, DOI 10.1007/978-1-4614-4286-8_3,

for some constant K independent of (t, x, y, u). For each control process $\nu \in \mathcal{U}$, we consider the controlled stochastic differential equation:

$$\mathrm{d}X_t = b(t, X_t, \nu_t)\mathrm{d}t + \sigma(t, X_t, \nu_t)\mathrm{d}W_t. \tag{3.3}$$

If the above equation has a unique solution X, for a given initial data, then the process X is called the controlled process, as its dynamics is driven by the action of the control process ν.

We shall be working with the following subclass of control processes:

$$\mathcal{U}_0 := \mathcal{U} \cap \mathbb{H}^2, \tag{3.4}$$

where $\mathbb{H}^2$ is the collection of all progressively measurable processes with finite $\mathbb{L}^2(\Omega \times [0,T))$–norm. Then, for every finite maturity $T' \leq T$, it follows from the above uniform Lipschitz condition on the coefficients b and σ that

$$\mathbb{E}\left[\int_0^{T'} \left(|b| + |\sigma|^2\right)(s, x, \nu_s)\mathrm{d}s\right] < \infty \quad \text{for all } \nu \in \mathcal{U}_0,\ x \in \mathbb{R}^d,$$

which guarantees the existence of a controlled process on the time interval $[0, T']$ for each given initial condition and control. The following result is an immediate consequence of Theorem 2.2.

Theorem 3.1. *Let $\nu \in \mathcal{U}_0$ be a control process and $\xi \in \mathbb{L}^2(\mathbb{P})$ be an $\mathcal{F}_0$–measurable random variable. Then, there exists a unique $\mathbb{F}$–adapted process X^ν satisfying* (3.3) *together with the initial condition $X_0^\nu = \xi$. Moreover, for every $T > 0$, there is a constant $C > 0$ such that*

$$\mathbb{E}\left[\sup_{0 \leq s \leq t} |X_s^\nu|^2\right] < C(1 + \mathbb{E}[|\xi|^2])\mathrm{e}^{Ct} \quad \textit{for all } t \in [0, T). \tag{3.5}$$

Gain Functional. Let

$$f, k : [0,T) \times \mathbb{R}^d \times U \longrightarrow \mathbb{R} \text{ and } g : \mathbb{R}^d \longrightarrow \mathbb{R}$$

be given functions. We assume that $f and k$ are continuous and $\|k^-\|_\infty < \infty$ (i.e., $\max(-k, 0)$ is uniformly bounded). Moreover, we assume that f and g satisfy the quadratic growth condition:

$$|f(t, x, u)| + |g(x)| \leq K(1 + |u| + |x|^2),$$

for some constant K independent of (t, x, u). We define the gain function J on $[0, T] \times \mathbb{R}^d \times \mathcal{U}$ by

$$J(t,x,\nu) := \mathbb{E}\left[\int_t^T \beta^\nu(t,s) f(s, X_s^{t,x,\nu}, \nu_s)\mathrm{d}s + \beta^\nu(t,T) g(X_T^{t,x,\nu})\mathbf{1}_{T<\infty}\right],$$

when this expression is meaningful, where

$$\beta^\nu(t,s) := \mathrm{e}^{-\int_t^s k(r, X_r^{t,x,\nu}, \nu_r)\mathrm{d}r},$$

and $\{X_s^{t,x,\nu}, s \geq t\}$ is the solution of (3.3) with control process ν and initial condition $X_t^{t,x,\nu} = x$.

Admissible Control Processes. In the finite horizon case $T < \infty$, the quadratic growth condition on f and g together with the bound on k^- ensure, that $J(t,x,\nu)$ is well defined for all control process $\nu \in \mathcal{U}_0$. We then define the set of admissible controls in this case by $\mathcal{U}_0$.

More attention is needed for the infinite horizon case. In particular, the discount term k needs to play a role to ensure the finiteness of the integral. In this setting the largest set of admissible control processes is given by

$$\mathcal{U}_0 := \left\{\nu \in \mathcal{U} : \mathbb{E}\left[\int_0^\infty \beta^\nu(t,s)\left(1 + |X_s^{t,x,\nu}|^2 + |\nu_s)|\right)\mathrm{d}s\right] < \infty \text{ for all } x\right\} \text{ when } T = \infty.$$

The Stochastic Control Problem. Consider the optimization problem

$$V(t,x) := \sup_{\nu \in \mathcal{U}_0} J(t,x,\nu) \text{ for } (t,x) \in \mathbf{S}.$$

Our main concern is to describe the local behavior of the value function V by means of the so-called *dynamic programming equation*, or *Hamilton-Jacobi-Bellman* equation. We continue with some remarks.

Remark 3.2. (i) If $V(t,x) = J(t,x,\hat{\nu}_{t,x})$, we call $\hat{\nu}_{t,x}$ an *optimal control* for the problem $V(t,x)$.

(ii) The following are some interesting subsets of controls :

- A process $\nu \in \mathcal{U}_0$ which is adapted to the natural filtration $\mathbb{F}^X$ of the associated state process is called *feedback control.*
- A process $\nu \in \mathcal{U}_0$ which can be written in the form $\nu_s = \tilde{u}(s, X_s)$ for some measurable map $\tilde{u}$ from $[0,T] \times \mathbb{R}^d$ into U, is called *Markovian control*; notice that any Markovian control is a feedback control.
- The deterministic processes of $\mathcal{U}_0$ are called *open loop controls.*

(iii) Suppose that $T < \infty$, and let (Y, Z) be the controlled processes defined by

$$\mathrm{d}Y_s = Z_s f(s, X_s, \nu_s)\mathrm{d}s \text{ and } \mathrm{d}Z_s = -Z_s k(s, X_s, \nu_s)\mathrm{d}s,$$

and define the augmented state process $\bar{X} := (X, Y, Z)$. Then, the above value function V can be written in the form:

$$V(t,x) = \bar{V}(t,x,0,1),$$

where $\bar{x} = (x, y, z)$ is some initial data for the augmented state process $\bar{X}$,

$$\bar{V}(t,\bar{x}) := \mathbb{E}_{t,\bar{x}}\left[\bar{g}(\bar{X}_T)\right] \text{ and } \bar{g}(x,y,z) := y + g(x)z .$$

Hence, the stochastic control problem V can be reduced without loss of generality to the case where $f = k \equiv 0$. We shall appeal to this reduced form whenever convenient for the exposition.

(iv) Consider the case $T < \infty$. In view of the previous remark, we may assume without loss of generality that $f = k = 0$. Consider the value function

$$\tilde{V}(t,x) := \sup_{\nu \in \mathcal{U}_t} \mathbb{E}\left[g(X_T^{t,x,\nu})\right], \tag{3.6}$$

differing from V by the restriction of the control processes to

$$\mathcal{U}_t := \{\nu \in \mathcal{U}_0 : \nu \text{ independent of } \mathcal{F}_t\}. \tag{3.7}$$

Since $\mathcal{U}_t \subset \mathcal{U}_0$, it is obvious that $\tilde{V} \le V$. We claim that

$$\tilde{V} = V, \tag{3.8}$$

so that both problems are indeed equivalent. To see this, fix $(t,x) \in \mathbf{S}$ and $\nu \in \mathcal{U}_0$. Then, ν can be written as a measurable function of the canonical process $\nu((\omega_s)_{0\le s\le t}, (\omega_s - \omega_t)_{t\le s\le T})$, where, for fixed $(\omega_s)_{0\le s\le t}$, the map $\nu_{(\omega_s)_{0\le s\le t}}$: $(\omega_s - \omega_t)_{t\le s\le T} \mapsto \nu((\omega_s)_{0\le s\le t}, (\omega_s - \omega_t)_{t\le s\le T})$ can be viewed as a control independent on $\mathcal{F}_t$. Using the independence of the increments of the Brownian motion, together with Fubini's lemma, it thus follows that

$$\begin{aligned} J(t,x;\nu) &= \int \mathbb{E}\left[g(X_T^{t,x,\nu_{(\omega_s)_{0\le s\le t}}})\right] d\mathbb{P}((\omega_s)_{0\le s\le t}) \\ &\le \int \tilde{V}(t,x) d\mathbb{P}((\omega_s)_{0\le s\le t}) \;=\; \tilde{V}(t,x). \end{aligned}$$

By arbitrariness of $\nu \in \mathcal{U}_0$, this implies that $\tilde{V}(t,x) \ge V(t,x)$.

(v) The extension of the previous remark to the infinite horizon case is also immediate.

3.2 The Dynamic Programming Principle

3.2.1 A Weak Dynamic Programming Principle

The dynamic programming principle is the main tool in the theory of stochastic control. In these notes, we shall prove rigorously a weak version of the dynamic programming which will be sufficient for the derivation of the dynamic programming equation. We denote:

$$V_*(t,x) := \liminf_{(t',x')\to(t,x)} V(t',x') \text{ and } V^*(t,x) := \limsup_{(t',x')\to(t,x)} V(t',x'),$$

for all $(t,x) \in \bar{\mathbf{S}}$. The functions V_* and V^* are called the lower and upper semicontinuous envelope, respectively. Clearly they are finite whenever the function V is locally bounded. We also recall the subset of controls $\mathcal{U}_t$ introduced in (3.7) above.

Theorem 3.3. *Assume that V is locally bounded. Let $(t,x) \in \mathbf{S}$ be fixed. Let $\{\theta^\nu,\ \nu \in \mathcal{U}_t\}$ be a family of finite stopping times independent of $\mathcal{F}_t$ with values in $[t,T]$. Then:*

$$V(t,x) \le \sup_{\nu\in\mathcal{U}_t} \mathbb{E}\left[\int_t^{\theta^\nu} \beta^\nu(t,s) f(s, X_s^{t,x,\nu}, \nu_s)\mathrm{d}s + \beta^\nu(t,\theta^\nu)V^*(\theta^\nu, X_{\theta^\nu}^{t,x,\nu})\right].$$

Assume further that g is lower semicontinuous and $X_{t,x}^\nu \mathbf{1}_{[t,\theta^\nu]}$ is $\mathbb{L}^\infty$–bounded for all $\nu \in \mathcal{U}_t$. Then

$$V(t,x) \ge \sup_{\nu\in\mathcal{U}_t} \mathbb{E}\left[\int_t^{\theta^\nu} \beta^\nu(t,s) f(s, X_s^{t,x,\nu}, \nu_s)\mathrm{d}s + \beta^\nu(t,\theta^\nu)V_*(\theta^\nu, X_{\theta^\nu}^{t,x,\nu})\right].$$

We shall provide an intuitive justification of this result after the following comments. A rigorous proof is reported in Sect. 3.2.2.

Remark. The possible dependence of θ on the central ν allows for more flexible choices of the stopping times and will be essential for our localization arguments in the rest of these lectures. In particular, the boundedness condition of $X_{t,x}^\nu \Pi_{[t,\theta^\nu]}$ will be automatically satisfied by localization.

Remark. (i) If V is continuous, then $V = V_* = V^*$, and the above weak dynamic programming principle reduces to the classical dynamic programming principle:

$$V(t,x) = \sup_{\nu\in\mathcal{U}} \mathbb{E}_{t,x}\left[\int_t^{\theta} \beta(t,s) f(s, X_s, \nu_s)\mathrm{d}s + \beta(t,\theta)V(\theta, X_\theta)\right]. \tag{3.9}$$

(ii) In the discrete-time framework, the dynamic programming principle (3.9) can be stated as follows :

$$V(t,x) = \sup_{u\in U} \mathbb{E}_{t,x}\left[f(t,X_t,u) + \mathrm{e}^{-k(t+1,X_{t+1},u)}V(t+1,X_{t+1})\right].$$

Observe that the supremum is now taken over the subset U of the finite dimensional space R^k. Hence, the dynamic programming principle allows to reduce the initial maximization problem, over the subset $\mathcal{U}$ of the infinite dimensional set of $\mathbb{R}^k-$valued processes, into a finite dimensional maximization problem. However, we are still facing an infinite dimensional problem since the dynamic programming principle relates the value function at time t to the value function at time $t+1$.

(iii) In the context of the above discrete-time framework with finite horizon $T < \infty$, notice that the dynamic programming principle suggests the following backward algorithm to compute V as well as the associated optimal strategy (when it exists). Since $V(T,\cdot) = g$ is known, the above dynamic programming principle can be applied recursively in order to deduce the value function $V(t,x)$ for every t.

(iv) In the continuous time setting, there is no obvious counterpart to the above backward algorithm. But, as the stopping time θ approaches t, the above dynamic programming principle implies a special local behavior for the value function V. When V is known to be smooth, this will be obtained by means of Itô's formula.

(v) It is usually very difficult to determine *a priori* the regularity of V. The situation is even worse since there are many counter-examples showing that the value function V can not be expected to be smooth in general; see Sect. 3.4. This problem is solved by appealing to the notion of viscosity solutions, which provides a weak local characterization of the value function V, and leads to the so-called dynamic programming equation or Hamilton-Jacobi-Bellman equation.

(vi) Once the local behavior of the value function is characterized, we are faced to the important uniqueness issue, which implies that V is completely characterized by its local behavior together with some convenient boundary condition.

Intuitive Justification of (3.9). Let us assume that V is continuous. In particular, V is measurable and $V = V_* = V^*$. Let $\tilde{V}(t,x)$ denote the right-hand side of (3.9).

By the tower property of the conditional expectation operator, it is easily checked that

$$J(t,x,\nu) = \mathbb{E}_{t,x}\left[\int_t^\theta \beta(t,s)f(s,X_s,\nu_s)\mathrm{d}s + \beta(t,\theta)J(\theta,X_\theta,\nu)\right].$$

Since $J(\theta,X_\theta,\nu) \le V(\theta,X_\theta)$, this proves that $V \le \tilde{V}$. To prove the reverse inequality, let $\mu \in \mathcal{U}$ and $\varepsilon > 0$ be fixed, and consider an $\varepsilon-$optimal control ν^ε for the problem $V(\theta,X_\theta)$, i.e.,

$$J(\theta, X_\theta, \nu^\varepsilon) \geq V(\theta, X_\theta) - \varepsilon.$$

Clearly, one can choose $\nu^\varepsilon = \mu$ on the stochastic interval $[t, \theta]$. Then

$$V(t,x) \geq J(t,x,\nu^\varepsilon) = \mathbb{E}_{t,x}\left[\int_t^\theta \beta(t,s) f(s, X_s, \mu_s) ds + \beta(t,\theta) J(\theta, X_\theta, \nu^\varepsilon)\right]$$

$$\geq \mathbb{E}_{t,x}\left[\int_t^\theta \beta(t,s) f(s, X_s, \mu_s) ds + \beta(t,\theta) V(\theta, X_\theta)\right] - \varepsilon\, \mathbb{E}_{t,x}[\beta(t,\theta)] .$$

This provides the required inequality by the arbitrariness of $\mu \in \mathcal{U}$ and $\varepsilon > 0$. □

Exercise. Where is the gap in the above sketch of the proof ?

Answer: The existence of the above control process ν^ε is not obvious because of crucial measurability problems!

3.2.2 Dynamic Programming Without Measurable Selection

In this section, we provide a rigorous proof of Theorem 3.3. Notice that, we have no information on whether V is measurable or not. Because of this, the right-hand side of the classical dynamic programming principle (3.9) is not even known to be well defined.

The formulation of Theorem 3.3 avoids this measurability problem since V_* and V^* are lower and upper semicontinuous, respectively, and therefore measurable. In addition, it allows to avoid the typically heavy technicalities related to measurable selection arguments needed for the proof of the classical dynamic programming principle (3.9) after a convenient relaxation of the control problem see, e.g., El Karoui and Jeanblanc [16].

Proof of Theorem 3.3. For simplicity, we consider the finite horizon case $T < \infty$, so that, without loss of generality, we assume $f = k = 0$; see Remark 3.2 (iii). The extension to the infinite horizon framework is immediate.

1. Let $\nu \in \mathcal{U}_t$ be arbitrary and set $\theta := \theta^\nu$. Then

$$\mathbb{E}\left[g\left(X_T^{t,x,\nu}\right) | \mathcal{F}_\theta\right](\omega) = J(\theta(\omega), X_\theta^{t,x,\nu}(\omega); \tilde{\nu}_\omega),$$

where $\tilde{\nu}_\omega$ is obtained from ν by freezing its trajectory up to the stopping time θ. Since, by definition, $J(\theta(\omega), X_\theta^{t,x,\nu}(\omega); \tilde{\nu}_\omega) \leq V^*(\theta(\omega), X_\theta^{t,x,\nu}(\omega))$, it follows from the tower property of conditional expectations that

$$\mathbb{E}\left[g\left(X_T^{t,x,\nu}\right)\right] = \mathbb{E}\left[\mathbb{E}\left[g\left(X_T^{t,x,\nu}\right) | \mathcal{F}_\theta\right]\right] \leq \mathbb{E}\left[V^*\left(\theta, X_\theta^{t,x,\nu}\right)\right],$$

which provides the second inequality of Theorem 3.3 by the arbitrariness of $\nu \in \mathcal{U}_t$.

2. Let $\varepsilon > 0$ be given, and consider an arbitrary function

$$\varphi : \mathbf{S} \longrightarrow \mathbb{R} \text{ such that } \varphi \text{ upper semicontinuous and } V \geq \varphi.$$

(a) There is a family $(\nu^{(s,y),\varepsilon})_{(s,y)\in\mathbf{S}} \subset \mathcal{U}_0$ such that:

$$\nu^{(s,y),\varepsilon} \in \mathcal{U}_s \text{ and } J(s,y;\nu^{(s,y),\varepsilon}) \geq V(s,y) - \varepsilon, \quad \text{for every}(s,y) \in \mathbf{S}. \tag{3.10}$$

Since g is lower semicontinuous and has quadratic growth, it follows from Theorem 3.1 that the function $(t',x') \mapsto J(t',x';\nu^{(s,y),\varepsilon})$ is lower semicontinuous, for fixed $(s,y) \in \mathbf{S}$. Together with the upper semicontinuity of φ, this implies that we may find a family $(r_{(s,y)})_{(s,y)\in\mathbf{S}}$ of positive scalars so that, for any $(s,y) \in \mathbf{S}$,

$$\varphi(s,y) - \varphi(t',x') \geq -\varepsilon \text{ and } J(s,y;\nu^{(s,y),\varepsilon}) - J(t',x';\nu^{(s,y),\varepsilon}) \leq \varepsilon$$
$$\text{for } (t',x') \in B(s,y;r_{(s,y)}), \tag{3.11}$$

where, for $r > 0$ and $(s,y) \in \mathbf{S}$,

$$B(s,y;r) := \left\{(t',x') \in \mathbf{S} \,:\, t' \in (s-r,s),\ |x'-y| < r\right\}. \tag{3.12}$$

Note that we do not use here balls of the usual form $B_r(s,y)$. The fact that $t' \leq s$ for $(t',x') \in B(s,y;r)$ will play an important role in Step 3. Clearly, $\left\{B(s,y;r) : (s,y) \in \mathbf{S},\ 0 < r \leq r_{(s,y)}\right\}$ forms an open covering of $[0,T) \times \mathbb{R}^d$. It then follows from the Lindelöf covering theorem, see, e.g., [15] Theorem 6.3 Chap. VIII, that we can find a countable sequence $(t_i,x_i,r_i)_{i\geq 1}$ of elements of $\mathbf{S} \times \mathbb{R}$, with $0 < r_i \leq r_{(t_i,x_i)}$ for all $i \geq 1$, such that $\mathbf{S} \subset \{T\} \times \mathbb{R}^d \cup (\cup_{i\geq 1} B(t_i,x_i;r_i))$. Set $A_0 := \{T\} \times \mathbb{R}^d$, $C_{-1} := \emptyset$, and define the sequence

$$A_{i+1} := B(t_{i+1},x_{i+1};r_{i+1}) \setminus C_i \text{ where } C_i := C_{i-1} \cup A_i,\ i \geq 0.$$

With this construction, it follows from (3.10) (3.11), together with the fact that $V \geq \varphi$, that the countable family $(A_i)_{i\geq 0}$ satisfies

$$(\theta, X_\theta^{t,x,\nu}) \in \cup_{i\geq 0} A_i \ \mathbb{P}-\text{a.s.},\ A_i \cap A_j = \emptyset \quad \text{for } i \neq j \in \mathbb{N},$$
$$\text{and } J(\cdot;\nu^{i,\varepsilon}) \geq \varphi - 3\varepsilon \text{ on } A_i \quad \text{for } i \geq 1, \tag{3.13}$$

where $\nu^{i,\varepsilon} := \nu^{(t_i,x_i),\varepsilon}$ for $i \geq 1$.

(b) We now prove the first inequality in Theorem 3.3. We fix $\nu \in \mathcal{U}_t$ and $\theta \in \mathcal{T}^t_{[t,T]}$. Set $A^n := \cup_{0\leq i\leq n} A_i$, $n \geq 1$. Given $\nu \in \mathcal{U}_t$, we define for $s \in [t,T]$:

$$\nu_s^{\varepsilon,n} := \mathbf{1}_{[t,\theta]}(s)\nu_s + \mathbf{1}_{(\theta,T]}(s)\left(\nu_s \mathbf{1}_{(A^n)^c}(\theta, X_\theta^{t,x,\nu}) + \sum_{i=1}^n \mathbf{1}_{A_i}(\theta, X_\theta^{t,x,\nu})\nu_s^{i,\varepsilon}\right).$$

Notice that $\{(\theta, X_\theta^{t,x,\nu}) \in A_i\} \in \mathcal{F}_\theta^t$, and therefore $\nu^{\varepsilon,n} \in \mathcal{U}_t$. By the definition of the neighborhood (3.12), notice that $\theta = \theta \wedge t_i \le t_i$ on $\{(\theta, X_{t,x}^\nu(\theta)) \in A_i\}$. Then, it follows from (3.13) that:

$$\begin{aligned}\mathbb{E}\left[g\left(X_T^{t,x,\nu^{\varepsilon,n}}\right)|\mathcal{F}_\theta\right]\mathbf{1}_{A^n}\left(\theta, X_\theta^{t,x,\nu}\right) &= V\left(T, X_T^{t,x,\nu^{\varepsilon,n}}\right)\mathbf{1}_{A_0}\left(\theta, X_\theta^{t,x,\nu}\right)\\ &\quad+\sum_{i=1}^n J(\theta, X_\theta^{t,x,\nu}, \nu^{i,\varepsilon})\mathbf{1}_{A_i}\left(\theta, X_\theta^{t,x,\nu}\right)\\ &\ge \sum_{i=0}^n \left(\varphi(\theta, X_\theta^{t,x,\nu} - 3\varepsilon\right)\mathbf{1}_{A_i}\left(\theta, X_\theta^{t,x,\nu}\right)\\ &= \left(\varphi(\theta, X_\theta^{t,x,\nu}) - 3\varepsilon\right)\mathbf{1}_{A^n}\left(\theta, X_\theta^{t,x,\nu}\right),\end{aligned}$$

which, by definition of V and the tower property of conditional expectations, implies

$$\begin{aligned}V(t,x) &\ge J(t,x,\nu^{\varepsilon,n})\\ &= \mathbb{E}\left[\mathbb{E}\left[g\left(X_T^{t,x,\nu^{\varepsilon,n}}\right)|\mathcal{F}_\theta\right]\right]\\ &\ge \mathbb{E}\left[\left(\varphi\left(\theta, X_\theta^{t,x,\nu}\right) - 3\varepsilon\right)\mathbf{1}_{A^n}\left(\theta, X_\theta^{t,x,\nu}\right)\right]\\ &\quad+\mathbb{E}\left[g\left(X_T^{t,x,\nu}\right)\mathbf{1}_{(A^n)^c}\left(\theta, X_\theta^{t,x,\nu}\right)\right].\end{aligned}$$

Since $g\left(X_T^{t,x,\nu}\right) \in \mathbb{L}^1$, it follows from the dominated convergence theorem that:

$$\begin{aligned}V(t,x) &\ge -3\varepsilon + \liminf_{n\to\infty}\mathbb{E}\left[\varphi(\theta, X_\theta^{t,x,\nu})\mathbf{1}_{A^n}\left(\theta, X_\theta^{t,x,\nu}\right)\right]\\ &= -3\varepsilon + \lim_{n\to\infty}\mathbb{E}\left[\varphi(\theta, X_\theta^{t,x,\nu})^+\mathbf{1}_{A^n}\left(\theta, X_\theta^{t,x,\nu}\right)\right]\\ &\quad-\lim_{n\to\infty}\mathbb{E}\left[\varphi(\theta, X_\theta^{t,x,\nu})^-\mathbf{1}_{A^n}\left(\theta, X_\theta^{t,x,\nu}\right)\right]\\ &= -3\varepsilon + \mathbb{E}\left[\varphi(\theta, X_\theta^{t,x,\nu})\right],\end{aligned}$$

where the last equality follows from the left-hand side of (3.13) and from the monotone convergence theorem, due to the fact that either $\mathbb{E}\left[\varphi(\theta, X_\theta^{t,x,\nu})^+\right] < \infty$ or $\mathbb{E}\left[\varphi(\theta, X_\theta^{t,x,\nu})^-\right] < \infty$. By the arbitrariness of $\nu \in \mathcal{U}_t$ and $\varepsilon > 0$, this shows that

$$V(t,x) \ge \sup_{\nu\in\mathcal{U}_t}\mathbb{E}\left[\varphi(\theta, X_\theta^{t,x,\nu})\right]. \tag{3.14}$$

3. It remains to deduce the first inequality of Theorem 3.3 from (3.14). Fix $r > 0$. It follows from standard arguments, see, e.g., Lemma 3.5 in [34], that we can find a

sequence of continuous functions $(\varphi_n)_n$ such that $\varphi_n \le V_* \le V$ for all $n \ge 1$ and such that φ_n converges pointwise to V_* on $[0,T]\times B_r(0)$. Set $\phi_N := \min_{n\ge N}\varphi_n$ for $N \ge 1$ and observe that the sequence $(\phi_N)_N$ is non-decreasing and converges pointwise to V_* on $[0,T]\times B_r(0)$. By (3.14) and the monotone convergence theorem, we then obtain

$$V(t,x) \ge \lim_{N\to\infty}\mathbb{E}\left[\phi_N(\theta^\nu, X^\nu_{t,x}(\theta^\nu))\right] = \mathbb{E}\left[V_*(\theta^\nu, X^\nu_{t,x}(\theta^\nu))\right]. \qquad \square$$

3.3 The Dynamic Programming Equation

The dynamic programming equation is the infinitesimal counterpart of the dynamic programming principle. It is also widely called the *Hamilton-Jacobi-Bellman* equation. In this section, we shall derive it under strong smoothness assumptions on the value function. Let $\mathcal{S}_d$ be the set of all $d\times d$ symmetric matrices with real coefficients, and define the map $H: \mathbf{S}\times\mathbb{R}\times\mathbb{R}^d\times\mathcal{S}_d$ by

$$\begin{aligned}H(t,x,r,p,\gamma)\\ := \sup_{u\in U}\left\{-k(t,x,u)r + b(t,x,u)\cdot p + \frac{1}{2}\mathrm{Tr}[\sigma\sigma^{\mathrm{T}}(t,x,u)\gamma] + f(t,x,u)\right\}.\end{aligned}$$

We also need to introduce the linear second-order operator $\mathcal{L}^u$ associated to the controlled process $\{\beta^u(0,t)X^u_t, t\ge 0\}$ controlled by the constant control process u:

$$\begin{aligned}\mathcal{L}^u\varphi(t,x) &:= -k(t,x,u)\varphi(t,x) + b(t,x,u)\cdot D\varphi(t,x)\\ &\quad +\frac{1}{2}\mathrm{Tr}\left[\sigma\sigma^{\mathrm{T}}(t,x,u)D^2\varphi(t,x)\right],\end{aligned}$$

where D and D^2 denote the gradient and the Hessian operators with respect to the x variable. With this notation, we have by Itô's formula

$$\begin{aligned}\beta^\nu(0,s)\varphi(s,X^\nu_s) - \beta^\nu(0,t)\varphi(t,X^\nu_t) &= \int_t^s \beta^\nu(0,r)\left(\partial_t + \mathcal{L}^{\nu_r}\right)\varphi(r,X^\nu_r)\mathrm{d}r\\ &\quad + \int_t^s \beta^\nu(0,r)D\varphi(r,X^\nu_r)\cdot\sigma(r,X^\nu_r,\nu_r)\mathrm{d}W_r\end{aligned}$$

for every $s\ge t$ and smooth function $\varphi\in C^{1,2}([t,s],\mathbb{R}^d)$ and each admissible control process $\nu\in\mathcal{U}_0$.

Proposition 3.4. *Assume the value function $V\in C^{1,2}([0,T),\mathbb{R}^d)$, and let the coefficients $k(\cdot,\cdot,u)$ and $f(\cdot,\cdot,u)$ be continuous in (t,x) for all fixed $u\in U$ and k bounded from below. Then, V is a classical supersolution of the dynamic programming equation, i.e. for all $(t,x)\in\mathbf{S}$,*

$$-\partial_t V(t,x) - H\left(t,x,V(t,x),DV(t,x),D^2V(t,x)\right) \geq 0. \tag{3.15}$$

Proof. Let $(t,x) \in \mathbf{S}$ and $u \in U$ be fixed, and consider the constant control process $\nu = u$, together with the associated state process X with initial data $X_t = x$. For all $h > 0$, define the stopping time:

$$\theta_h := \inf\{s > t \ : \ (s-t, X_s - x) \notin [0,h) \times \alpha B\},$$

where $\alpha > 0$ is some given constant, and B denotes the unit ball of $\mathbb{R}^d$. Notice that $\theta_h \longrightarrow t$, $\mathbb{P}$–a.s. when $h \searrow 0$, and $\theta_h = t + h < T$ for $h \leq \bar{h}(\omega)$ sufficiently small.

1. From the first inequality of the dynamic programming principle, together with the continuity of V, it follows that

$$0 \leq \mathbb{E}_{t,x}\left[\beta^u(0,t)V(t,x) - \beta^u(0,\theta_h)V(\theta_h, X_{\theta_h}) - \int_t^{\theta_h} \beta^u(0,r)f(r,X_r,u)\mathrm{d}r\right]$$
$$= -\mathbb{E}_{t,x}\left[\int_t^{\theta_h} \beta^u(0,r)(\partial_t V + \mathcal{L}^{\cdot} V + f)(r,X_r,u)\mathrm{d}r\right]$$
$$-\mathbb{E}_{t,x}\left[\int_t^{\theta_h} \beta^u(0,r)DV(r,X_r)\cdot\sigma(r,X_r,u)\mathrm{d}W_r\right];$$

   the last equality follows from Itô's formula and uses the crucial smoothness assumption on V.
2. Observe that $\beta(0,r)DV(r,X_r)\cdot\sigma(r,X_r,u)$ is bounded on the stochastic interval $[t,\theta_h]$. Therefore, the second expectation on the right hand-side of the last inequality vanishes, and we obtain:

$$-\mathbb{E}_{t,x}\left[\frac{1}{h}\int_t^{\theta_h} \beta^u(0,r)(\partial_t V + \mathcal{L}^{\cdot} V + f)(r,X_r,u)\mathrm{d}r\right] \geq 0$$

   We now send h to zero. The a.s. convergence of the random value inside the expectation is easily obtained by the mean value theorem; recall that $\theta_h = h$ for sufficiently small $h > 0$. Since the random variable $h^{-1}\int_t^{\theta_h} \beta^u(0,r)(\mathcal{L}^{\cdot} V + f)(r,X_r,u)\mathrm{d}r$ is essentially bounded, uniformly in h, on the stochastic interval $[t,\theta_h]$, it follows from the dominated convergence theorem that

$$-\partial_t V(t,x) - \mathcal{L}^u V(t,x) - f(t,x,u) \geq 0.$$

   By the arbitrariness of $u \in U$, this provides the required claim. □

We next wish to show that V satisfies the nonlinear partial differential equation (3.15) with equality. This is a more technical result which can be proved by different methods. We shall report a proof, based on a contradiction argument, which provides more intuition on this result, although it might be slightly longer than the usual proof reported in standard textbooks.

Proposition 3.5. *Assume* $V \in C^{1,2}([0,T),\mathbb{R}^d)$ *and* $H(., V.DV, D^2V) < \infty$. *Assume further that* k *is bounded and the function* H *is upper semicontinuous. Then,* V *is a classical subsolution of the dynamic programming equation, i.e. for all* $(t,x) \in \mathbf{S}$

$$-\partial_t V(t,x) - H\left(t,x,V(t,x),DV(t,x),D^2V(t,x)\right) \le 0. \tag{3.16}$$

Proof. Let $(t_0,x_0) \in [0,T) \times \mathbb{R}^d$ be fixed, assume to the contrary that

$$\partial_t V(t_0,x_0) + H\left(t_0,x_0,V(t_0,x_0),DV(t_0,x_0),D^2V(t_0,x_0)\right) < 0, \tag{3.17}$$

and let us work towards a contradiction.

1. For a given parameter $\varepsilon > 0$, define the smooth function $\varphi \ge V$ by

$$\varphi(t,x) := V(t,x) + \varepsilon\left(|t-t_0|^2 + |x-x_0|^4\right).$$

Then

$$(V-\varphi)(t_0,x_0) = 0, \ \ (DV - D\varphi)(t_0,x_0) = 0, \ \ (\partial_t V - \partial_t\varphi)(t_0,x_0) = 0,$$
$$\text{and} \ \ (D^2V - D^2\varphi)(t_0,x_0) = 0,$$

and it follows from the continuity of H and (3.17) that:

$$h(t,x) := \left(\partial_t\varphi + H(.,\varphi,D\varphi,D^2\varphi)\right)(t,x) < 0 \ \text{ for } (t,x) \in \mathcal{N}_r, \tag{3.18}$$

for some sufficiently small parameter $r > 0$, where $\mathcal{N}_r := (t_0 - r, t_0 + r) \times rB(x_0) \subset [-r,T] \times \mathbb{R}^n$, and $B(x_0)$ is the open unit ball centered at x_0.

2. From the definition of φ, we have

$$-\eta := \max_{\partial\mathcal{N}_r}(V - \varphi) < 0. \tag{3.19}$$

For an arbitrary control process $\nu \in \mathcal{U}_{t_0}$, we define the stopping time

$$\theta^\nu := \inf\{t > t_0 : \left(t, X_t^{t_0,x_0,\nu}\right) \notin \mathcal{N}_r\},$$

and we observe that $\left(\theta^\nu, X_{\theta^\nu}^{t_0,x_0,\nu}\right) \in \partial\mathcal{N}_r$ by the pathwise continuity of the controlled process. Then, it follows from (3.19) that

$$\varphi\left(\theta^\nu, X_{\theta^\nu}^{t_0,x_0,\nu}\right) \ge \eta + V\left(\theta^\nu, X_{\theta^\nu}^{t_0,x_0,\nu}\right). \tag{3.20}$$

3. For notation simplicity, we set $\beta_s^\nu := \beta^\nu(t_0,s)$. Since $\beta_{t_0}^\nu = 1$, it follows from Itô's formula together with the boundedness of β (implied by the lower boundedness of k) that:

$$V(t_0, x_0) = \varphi(t_0, x_0)$$
$$= \mathbb{E}\Big[\beta^\nu_{\theta^\nu}\varphi(\theta^\nu, X^{t_0,x_0,\nu}_{\theta^\nu}) - \int_{t_0}^{\theta^\nu} \beta^\nu_s(\partial_t + \mathcal{L}^{\nu_s})\varphi(s, X^{t_0,x_0,\nu}_s)\mathrm{d}s\Big]$$
$$\geq \mathbb{E}\Big[\beta^\nu_{\theta^\nu}\varphi(\theta^\nu, X^{t_0,x_0,\nu}_{\theta^\nu}) + \int_{t_0}^{\theta^\nu} \beta^\nu_s\big(f(.,\nu_s) - h\big)(s, X^{t_0,x_0,\nu}_s)\mathrm{d}s\Big]$$

by the definition of h. Since $(s, X^{t_0,x_0,\nu}_s) \in \mathcal{N}_r$ on $[t_0, \theta^\nu)$, it follows from (3.18) and (3.20) that

$$V(t_0, x_0) \geq \eta\mathbb{E}\left[\beta^\nu_{\theta^\nu}\right] + \mathbb{E}\left[\int_{t_0}^{\theta^\nu} \beta^\nu_s f\left(s, X^{t_0,x_0,\nu}_s, \nu_s\right)\mathrm{d}s + \beta^\nu_{\theta^\nu} V\left(\theta^\nu, X^{t_0,x_0,\nu}_{\theta^\nu}\right)\right]$$
$$\geq \eta \mathrm{e}^{-r|k^+|_\infty} + \mathbb{E}\left[\int_{t_0}^{\theta^\nu} \beta^\nu_s f\left(s, X^{t_0,x_0,\nu}_s, \nu_s\right)\mathrm{d}s + \beta^\nu_{\theta^\nu} V\left(\theta^\nu, X^{t_0,x_0,\nu}_{\theta^\nu}\right)\right].$$

Since $\eta > 0$ does not depend on ν, it follows from the arbitrariness of $\nu \in \mathcal{U}_{t_0}$ and the continuity of V that the last inequality is in contradiction with the second inequality of the dynamic programming principle of Theorem 3.3. □

As a consequence of Propositions 3.4 and 3.5, we have the main result of this section:

Theorem 3.6. *Let the conditions of Propositions 3.5 and 3.4 hold. Then, the value function V solves the Hamilton-Jacobi-Bellman equation*

$$-\partial_t V - H\left(., V, DV, D^2V\right) = 0 \text{ on } \mathbf{S}. \tag{3.21}$$

3.4 On the Regularity of the Value Function

The purpose of this paragraph is to show that the value function should not be expected to be smooth in general. We start by proving the continuity of the value function under strong conditions; in particular, we require the set U in which the controls take values to be bounded. We then give a simple example in the deterministic framework where the value function is not smooth. Since it is well known that stochastic problems are "more regular" than deterministic ones, we also give an example of stochastic control problem whose value function is not smooth.

3.4.1 Continuity of the Value Function for Bounded Controls

For simplicity, we reduce the stochastic control problem to the case $f = k \equiv 0$, see Remark 3.2iii). We will also concentrate on the finite horizon case $T < \infty$.

The corresponding results in the infinite horizon case can be obtained by similar arguments. Our main concern, in this section, is to show the standard argument for proving the continuity of the value function. Therefore, the following results assume strong conditions on the coefficients of the model in order to simplify the proofs. We first start by examining the value function $V(t,\cdot)$ for fixed $t \in [0,T]$.

Proposition 3.7. *Let $f = k \equiv 0$, $T < \infty$, and assume that g is Lipschitz continuous. Then*

(i) *V is Lipschitz in x, uniformly in t.*
(ii) *Assume further that U is bounded. Then V is $\frac{1}{2}$–Hölder continuous in t, and there is a constant $C > 0$ such that*

$$\left|V(t,x) - V(t',x)\right| \leq C(1+|x|)\sqrt{|t-t'|};\quad t,t' \in [0,T],\ x \in \mathbb{R}^d.$$

Proof. (i) For $x, x' \in \mathbb{R}^d$ and $t \in [0,T)$, we first estimate that

$$\begin{aligned}\left|V(t,x) - V(t,x')\right| &\leq \sup_{\nu\in\mathcal{U}_0} \mathbb{E}\left|g\left(X_T^{t,x,\nu}\right) - g\left(X_T^{t,x',\nu}\right)\right| \\ &\leq \text{Const} \sup_{\nu\in\mathcal{U}_0} \mathbb{E}\left|X_T^{t,x,\nu} - X_T^{t,x',\nu}\right| \\ &\leq \text{Const}\,|x-x'|,\end{aligned}$$

where we used the Lipschitz continuity of g together with the flow estimates of Theorem 2.4, and the fact that the coefficients b and σ are Lipschitz in x uniformly in (t,u). This completes the proof of the Lipschitz property of the value function V.

(ii) To prove the Hölder continuity in t, we shall use the dynamic programming principle.

(1) We first make the following important observation. A careful review of the proof of Theorem 3.3 reveals that, whenever the stopping times θ^ν are constant (i.e., deterministic), the dynamic programming principle holds true with the semicontinuous envelopes taken only with respect to the x–variable. Since V was shown to be continuous in the first part of this proof, we deduce that

$$V(t,x) = \sup_{\nu\in\mathcal{U}_0} \mathbb{E}\left[V\left(t', X_{t'}^{t,x,\nu}\right)\right] \tag{3.22}$$

for all $x \in \mathbb{R}^d$, $t < t' \in [0,T]$.

(2) Fix $x \in \mathbb{R}^d$, $t < t' \in [0,T]$. By the dynamic programming principle (3.22), we have

$$
\begin{aligned}
|V(t,x) - V(t',x)| &= \left| \sup_{\nu \in \mathcal{U}_0} \mathbb{E}\left[V\left(t', X_{t'}^{t,x,\nu}\right)\right] - V(t',x) \right| \\
&\leq \sup_{\nu \in \mathcal{U}_0} \mathbb{E}\left|V\left(t', X_{t'}^{t,x,\nu}\right) - V(t',x)\right|.
\end{aligned}
$$

By the Lipschitz continuity of $V(s,\cdot)$ established in the first part of this proof, we see that

$$
|V(t,x) - V(t',x)| \leq \text{Const} \sup_{\nu \in \mathcal{U}_0} \mathbb{E}\left|X_{t'}^{t,x,\nu} - x\right|. \tag{3.23}
$$

We shall now prove that

$$
\sup_{\nu \in \mathcal{U}} \mathbb{E}\left|X_{t'}^{t,x,\nu} - x\right| \leq \text{Const}\,(1+|x|)|t-t'|^{1/2}, \tag{3.24}
$$

which provides the required $(1/2)-$Hölder continuity in view of (3.23). By definition of the process X, and assuming $t < t'$, we have

$$
\begin{aligned}
\mathbb{E}\left|X_{t'}^{t,x,\nu} - x\right|^2 &= \mathbb{E}\left|\int_t^{t'} b(r, X_r, \nu_r)\mathrm{d}r + \int_t^{t'} \sigma(r, X_r, \nu_r)\mathrm{d}W_r\right|^2 \\
&\leq \text{Const}\,\mathbb{E}\left[\int_t^{t'} |h(r, X_r, \nu_r)|^2\,\mathrm{d}r\right]
\end{aligned}
$$

where $h := [b^2 + \sigma^2]^{1/2}$. Since h is Lipschitz continuous in (t,x,u) and has quadratic growth in x and u, this provides

$$
\begin{aligned}
\mathbb{E}\left|X_{t'}^{t,x,\nu} - x\right|^2 \leq \text{Const} \Bigg(&\int_t^{t'} (1 + |x|^2 + |\nu_r|^2)\mathrm{d}r \\
&+ \int_t^{t'} \mathbb{E}\left|X_r^{t,x,\nu} - x\right|^2 \mathrm{d}r \Bigg).
\end{aligned}
$$

Since the control process ν is uniformly bounded, we obtain by the Gronwall lemma the estimate:

$$
\mathbb{E}\left|X_{t'}^{t,x,\nu} - x\right|^2 \leq \text{Const}\,(1+|x|)|t'-t|, \tag{3.25}
$$

where the constant does not depend on the control ν. □

Remark 3.8. When f and/or k are non-zero, the conditions required on f and k in order to obtain the $(1/2)-$Hölder continuity of the value function can be deduced from the reduction of Remark 3.2(iii).

Remark 3.9. Further regularity results can be proved for the value function under convenient conditions. Typically, one can prove that $\mathcal{L}^u V$ exists in the generalized sense, for all $u \in U$. This implies immediately that the result of Proposition 3.5 holds in the generalized sense. More technicalities are needed in order to derive the result of Proposition 3.4 in the generalized sense. We refer to [20], Sect. IV.10, for a discussion of this issue and to Krylov [27] for the technical proofs.

3.4.2 A Deterministic Control Problem with Non-smooth Value Function

Let $\sigma \equiv 0$, $b(x,u) = u$, $U = [-1,1]$, and $n = 1$. The controlled state is then the one-dimensional deterministic process defined by

$$X_s = X_t + \int_t^s \nu_t \mathrm{d}t \quad \text{for } 0 \le t \le s \le T .$$

Consider the deterministic control problem

$$V(t,x) := \sup_{\nu \in \mathcal{U}} (X_T)^2.$$

The dynamic programming equation corresponding to this problem is:

$$-\partial_t V - |D\nu| = 0.$$

The value function of this problem is easily seen to be given by

$$V(t,x) = \begin{cases} (x+T-t)^2 & \text{for } x \ge 0 \text{ with optimal control } \hat{u} = 1, \\ (x-T+t)^2 & \text{for } x \le 0 \text{ with optimal control } \hat{u} = -1. \end{cases}$$

This function is continuous. However, a direct computation shows that it is not differentiable at $x = 0$.

3.4.3 A Stochastic Control Problem with Non-smooth Value Function

Let $U = \mathbb{R}$ and the controlled process X^ν be the scalar process defined by the dynamics:

$$\mathrm{d}X_t^\nu = \nu_t \mathrm{d}W_t,$$

where W is a scalar Brownian motion. Then, for any $\nu \in \mathcal{U}_0$, the process X^ν is a martingale. Let g be a function defined on $\mathbb{R}$ with linear growth $|g(x)| \le C(1+|x|)$ for some constant $C > 0$. Then $g(X_T^\nu)$ is integrable for all $T \ge 0$. Consider the stochastic control problem

$$V(t,x) := \sup_{\nu \in \mathcal{U}_0} \mathbb{E}_{t,x}\left[g(X_T^\nu)\right].$$

Let us assume that V is smooth, and work towards a contradiction.

1. If V is $C^{1,2}([0,T),\mathbb{R})$, then it follows from Proposition 3.4 that V is a supersolution of the dynamic programming equation:

$$-\partial_t V - \frac{1}{2}u^2 D^2 V \ge 0 \quad \text{for all } u \in \mathbb{R},$$

   and all $(t,x) \in [0,T) \times \mathbb{R}$. By sending u to infinity, it follows that

$$V(t,\cdot) \text{ is concave for all } t \in [0,T). \tag{3.26}$$

2. Notice that $V(t,x) \ge \mathbb{E}_{t,x}\left[g(X_T^0)\right] = g(x)$. Then, it follows from (3.26) that:

$$V(t,x) \ \ge\ g^{\text{conc}}(x) \quad \text{for all } (t,x) \in [0,T) \times \mathbb{R}, \tag{3.27}$$

   where g^{conc} denotes the concave envelope of g, i.e., the smallest concave majorant of g. If $g^{\text{conc}} = \infty$, this already proves that $V = \infty$. We then continue in the case that $g^{\text{conc}} < \infty$.
3. Since $g \le g^{\text{conc}}$, we see that

$$V(t,x) \ :=\ \sup_{\nu \in \mathcal{U}_0} \mathbb{E}_{t,x}\left[g(X_T^\nu)\right] \ \le\ \sup_{\nu \in \mathcal{U}_0} \mathbb{E}_{t,x}\left[g^{\text{conc}}(X_T^\nu)\right]. \ \le\ g^{\text{conc}}(x),$$

   where the last equality follows from the Jensen inequality together with the martingale property of the controlled process X^ν. In view of (3.27), we have then proved that

$$\begin{aligned} & V \ \in\ C^{1,2}([0,T),\mathbb{R}) \\ & \quad \Longrightarrow\ V(t,x) \ =\ g^{\text{conc}}(x) \quad \text{for all } (t,x) \in [0,T) \times \mathbb{R}. \end{aligned}$$

   Now recall that this implication holds for any arbitrary function g with linear growth. We then obtain a contradiction whenever the function g^{conc} is not $C^2(\mathbb{R})$. Hence,

$$g^{\text{conc}} \ \notin\ C^2(\mathbb{R}) \Longrightarrow V \ \notin\ C^{1,2}([0,T),\mathbb{R}^2).$$

Chapter 4
Optimal Stopping and Dynamic Programming

As in the previous chapter, we assume here that the filtration $\mathbb{F}$ is defined as the $\mathbb{P}-$augmentation of the canonical filtration of the Brownian motion W defined on the probability space $(\Omega, \mathcal{F}, \mathbb{P})$.

Our objective is to derive similar results, as those obtained in the previous chapter for standard stochastic control problems, in the context of optimal stopping problems. We will then first start with the formulation of optimal stopping problems, then the corresponding dynamic programming principle, and dynamic programming equation.

4.1 Optimal Stopping Problems

For $0 \leq t \leq T < \infty$, we denote by $\mathcal{T}_{[t,T]}$ the collection of all $\mathbb{F}-$stopping times with values in $[t, T]$. We also recall the notation $\mathbf{S} := [0, T) \times \mathbb{R}^n$ for the parabolic state space of the underlying state process X defined by the stochastic differential equation:

$$dX_t = b(t, X_t)dt + \sigma(t, X_t)dW_t, \tag{4.1}$$

where b and σ are defined on $\bar{\mathbf{S}}$ and take values in $\mathbb{R}^n$ and $\mathcal{S}_n$, respectively. We assume that b and σ satisfy the usual Lipschitz and linear growth conditions so that the above SDE has a unique strong solution satisfying the integrability proved in Theorem 2.2.

The infinitesimal generator of the Markovian diffusion process X is denoted by

$$\mathcal{A}\varphi := b \cdot D\varphi + \frac{1}{2}\text{Tr}\left[\sigma\sigma^{\text{T}} D^2\varphi\right].$$

N. Touzi, *Optimal Stochastic Control, Stochastic Target Problems, and Backward SDE*, Fields Institute Monographs 29, DOI 10.1007/978-1-4614-4286-8_4,

Let g be a continuous function from $\mathbb{R}^n$ to $\mathbb{R}$ and assume that

$$\mathbb{E}\left[\sup_{0\le t\le T}|g(X_t)|\right] < \infty. \tag{4.2}$$

For instance, if g has linear growth, the previous integrability condition is automatically satisfied. Under this condition, the criterion

$$J(t,x,\tau) := \mathbb{E}\left[g\left(X_\tau^{t,x}\right)\right] \tag{4.3}$$

is well defined for all $(t,x) \in \mathbf{S}$ and $\tau \in \mathcal{T}_{[t,T]}$. Here, $X^{t,x}$ denotes the unique strong solution of (4.1) with initial condition $X_t^{t,x} = x$.

The optimal stopping problem is now defined by:

$$V(t,x) := \sup_{\tau\in\mathcal{T}_{[t,T]}} J(t,x,\tau) \quad \text{for all } (t,x) \in \mathbf{S}. \tag{4.4}$$

A stopping time $\hat{\tau} \in \mathcal{T}_{[t,T]}$ is called an optimal stopping rule if $V(t,x) = J(t,x,\hat{\tau})$.

The set

$$\mathcal{S} := \{(t,x) : V(t,x) = g(x)\} \tag{4.5}$$

is called the *stopping region* and is of particular interest: whenever the state is in this region, it is optimal to stop immediately. Its complement $\mathcal{S}^c$ is called the *continuation region*.

Remark 4.1. As in the previous chapter we could have allowed for the infinite horizon $T \le \infty$, and we could have considered the apparently more general criterion

$$\bar{V}(t,x) := \sup_{\tau\in\mathcal{T}_{[t,T]}} \mathbb{E}\left[\int_t^\tau \beta(t,s)f(s,X_s)\mathrm{d}s + \beta(t,\tau)g\left(X_\tau^{t,x}\right)\mathbf{1}_{\tau<\infty}\right],$$

with

$$\beta(t,s) := \mathrm{e}^{-\int_t^s k(s,X_s)\mathrm{d}s} \quad \text{for } 0 \le t \le s < T.$$

In the finite horizon case, this problem may be reduce to the context of (4.4) by introducing the additional states

$$Y_t := Y_0 + \int_0^t \beta_s f(s,X_s)\mathrm{d}s,$$

$$Z_t := Z_0 + \int_0^t Z_s k(s,X_s)\mathrm{d}s.$$

Also, the methodology which will be developed for the problem (4.4) naturally extends to the infinite horizon case.

Remark 4.2. Consider the subset of stopping rules:

$$\mathcal{T}^t_{[t,T]} := \left\{\tau \in \mathcal{T}_{[t,T]} : \ \tau \text{ independent of } \mathcal{F}_t\right\}. \tag{4.6}$$

By a similar argument as in Remark 3.2iv), we can see that the maximization in the optimal stopping problem (4.4) can be restricted to this subset, i.e.,

$$V(t,x) := \sup_{\tau \in \mathcal{T}^t_{[t,T]}} J(t,x,\tau) \quad \text{for all } (t,x) \in \mathbf{S}. \tag{4.7}$$

4.2 The Dynamic Programming Principle

The proof of the dynamic programming principle for optimal stopping problems is easier than in the context of stochastic control problems of the previous chapter. The reader may consult the excellent exposition in the book of Karatzas and Shreve [24], Appendix D, where the following dynamic programming principle is proved:

$$V(t,x) = \sup_{\tau \in \mathcal{T}^t_{[t,T]}} \mathbb{E}\left[\mathbf{1}_{\{\tau<\theta\}} g(X^{t,x}_\tau) + \mathbf{1}_{\{\tau \geq \theta\}} V(\theta, X^{t,x}_\theta)\right], \tag{4.8}$$

for all $(t,x) \in \mathbf{S}$ and $\tau \in \mathcal{T}_{[t,T]}$. In particular, the proof in the previous reference does not require any heavy measurable selection, and is essentially based on the supermartingale nature of the so-called Snell envelope process. Moreover, we observe that it does not require any Markovian property of the underlying state process.

We report here a different proof in the spirit of the weak dynamic programming principle for stochastic control problems proved in the previous chapter. The subsequent argument is specific to our Markovian framework and, in this sense, is weaker than the classical dynamic programming principle. However, the combination of the arguments of this chapter with those of the previous chapter allows to derive a dynamic programming principle for mixed stochastic control and stopping problems.

The following claim will make use of the subset $\mathcal{T}^t_{[t,T]}$, introduced in (4.6), of all stopping times in $\mathcal{T}_{[t,T]}$ which are independent of $\mathcal{F}_t$, and the notations:

$$V_*(t,x) := \liminf_{(t',x') \to (t,x)} V(t',x') \text{ and } V^*(t,x) := \limsup_{(t',x') \to (t,x)} V(t',x'),$$

for all $(t,x) \in \bar{\mathbf{S}}$. We recall that V_* and V^* are the lower-and upper-semicontinuous envelopes of V, and that $V_* = V^* = V$ whenever V is continuous.

Theorem 4.3. *Assume that V is locally bounded. For $(t,x) \in \mathbf{S}$, let $\theta \in \mathcal{T}^t_{[t,T]}$ be a stopping time such that $X^{t,x}_\theta$ is $\mathbb{L}^\infty(\mathbb{P})-$ bounded. Then*

$$V(t,x) \leq \sup_{\tau \in \mathcal{T}^t_{[t,T]}} \mathbb{E}\left[\mathbf{1}_{\{\tau<\theta\}} g(X^{t,x}_\tau) + \mathbf{1}_{\{\tau\geq\theta\}} V^*(\theta, X^{t,x}_\theta)\right], \tag{4.9}$$

$$V(t,x) \geq \sup_{\tau \in \mathcal{T}^t_{[t,T]}} \mathbb{E}\left[\mathbf{1}_{\{\tau<\theta\}} g(X^{t,x}_\tau) + \mathbf{1}_{\{\tau\geq\theta\}} V_*(\theta, X^{t,x}_\theta))\right]. \tag{4.10}$$

Proof. Inequality (4.9) follows immediately from the tower property and the fact that $J \leq V^*$.

We next prove inequality (4.10) with V_* replaced by an arbitrary function

$$\varphi : \mathbf{S} \longrightarrow \mathbb{R} \text{ such that } \varphi \text{ is upper-semicontinuous and } V \geq \varphi,$$

which implies (4.10) by the same argument as in Step 3 of the proof of Theorem 3.3.

Arguing as in Step 2 of the proof of Theorem 3.3, we first observe that, for every $\varepsilon > 0$, we can find a countable family $\bar{A}_i \subset (t_i - r_i, t_i] \times A_i \subset \mathbf{S}$, together with a sequence of stopping times $\tau^{i,\varepsilon}$ in $\mathcal{T}^{t_i}_{[t_i,T]}$, $i \geq 1$, satisfying $\bar{A}_0 = \{T\} \times \mathbb{R}^d$ and

$$\cup_{i\geq 0}\bar{A}_i = \mathbf{S},\ \bar{A}_i \cap \bar{A}_j = \emptyset \text{ for } i \neq j \in \mathbb{N},\ \bar{J}(\cdot;\tau^{i,\varepsilon}) \geq \varphi - 3\varepsilon \text{ on } \bar{A}_i \text{ for } i \geq 1. \tag{4.11}$$

Set $\bar{A}^n := \cup_{i\leq n}\bar{A}_i$, $n \geq 1$. Given two stopping times $\theta, \tau \in \mathcal{T}^t_{[t,T]}$, it is easily checked that

$$\tau^{n,\varepsilon} := \tau\mathbf{1}_{\{\tau<\theta\}} + \mathbf{1}_{\{\tau\geq\theta\}}\left(T\mathbf{1}_{(\bar{A}^n)^c}\left(\theta, X^{t,x}_\theta\right) + \sum_{i=1}^n \tau^{i,\varepsilon}\mathbf{1}_{\bar{A}_i}\left(\theta, X^{t,x}_\theta\right)\right)$$

defines a stopping time in $\mathcal{T}^t_{[t,T]}$. We then deduce from the tower property and (4.11) that

$$\begin{aligned}
\bar{V}(t,x) &\geq \bar{J}(t,x;\tau^{n,\varepsilon}) \\
&\geq \mathbb{E}\left[g\left(X^{t,x}_\tau\right)\mathbf{1}_{\{\tau<\theta\}} + \mathbf{1}_{\{\tau\geq\theta\}}\left(\varphi(\theta, X^{t,x}_\theta) - 3\varepsilon\right)\mathbf{1}_{\bar{A}^n}(\theta, X^{t,x}_\theta)\right] \\
&\quad + \mathbb{E}\left[\mathbf{1}_{\{\tau\geq\theta\}} g(X^{t,x}_T)\mathbf{1}_{(\bar{A}^n)^c}(\theta, X^{t,x}_\theta)\right].
\end{aligned}$$

By sending $n \to \infty$ and arguing as in the end of Step 2 of the proof of Theorem 3.3, we deduce that

$$\bar{V}(t,x) \geq \mathbb{E}\left[g\left(X^{t,x}_\tau\right)\mathbf{1}_{\{\tau<\theta\}} + \mathbf{1}_{\{\tau\geq\theta\}}\varphi(\theta, X^{t,x}_\theta)\right] - 3\varepsilon,$$

and the result follows from the arbitrariness of $\varepsilon > 0$ and $\tau \in \mathcal{T}^t_{[t,T]}$. □

Remark 4.4. In the context of the optimal stopping $\bar{V}$ introduced in Remark 4.1, denoting $\beta^s_t := \beta(t,s)$, the dynamic programming principle of Theorem 4.3 translates to

$$\bar{V}(t,x) \le \sup_{\tau\in\mathcal{T}^t_{[t,T]}} \mathbb{E}\left[\int_t^{\theta\wedge\tau} \beta^s_t f(s,X^{t,x}_s)ds + \beta^{\theta\wedge\tau}_t\Big(\mathbf{1}_{\tau<\theta} g(X^{t,x}_\tau) + \mathbf{1}_{\tau\ge\theta}\bar{V}^*(\theta,X^{t,x}_\theta)\Big)\right],$$

$$\bar{V}(t,x) \ge \sup_{\tau\in\mathcal{T}^t_{[t,T]}} \mathbb{E}\left[\int_t^{\theta\wedge\tau} \beta^s_t f(s,X^{t,x}_s)ds + \beta^{\theta\wedge\tau}_t\Big(\mathbf{1}_{\tau<\theta} g(X^{t,x}_\tau) + \mathbf{1}_{\tau\ge\theta}\bar{V}_*(\theta,X^{t,x}_\theta)\Big)\right],$$

and suitable conditions inherited from the reduction of Remark 4.1. In this setting, the derivation of the dynamic programming equation, in the subsequent sections, involves the linear second order differential operator

$$\mathcal{L}\varphi := -k\varphi + \mathcal{A}\varphi = -k\varphi + b\cdot D\varphi + \frac{1}{2}\mathrm{Tr}\big[\sigma\sigma^{\mathrm{T}}D^2\varphi\big]. \tag{4.12}$$

4.3 The Dynamic Programming Equation

In this section, we explore the infinitesimal counterpart of the dynamic programming principle of Theorem 4.3, when the value function V is a priori known to be smooth. The smoothness that will be required in this chapter must be so that Itô's formula applies to V. In particular, V is continuous, and the dynamic programming principle of Theorem 4.3 reduces to the classical dynamic programming principle (4.8).

Loosely speaking, the following dynamic programming equation says the following:

- In the stopping region $\mathcal{S}$ defined in (4.5), continuation is sub-optimal, and therefore the linear PDE must hold with inequality in such a way that the value function is a submartingale.
- In the continuation region $\mathcal{S}^c$, it is optimal to delay the stopping decision after some small moment, and therefore the value function must solve a linear PDE as in Chap. 2.

Theorem 4.5. *Assume that $V \in C^{1,2}([0,T),\mathbb{R}^n)$, and let $g:\mathbb{R}^n \longrightarrow \mathbb{R}$ be continuous. Then V solves the obstacle problem:*

$$\min\{-(\partial_t+\mathcal{A})V,\ V-g\} = 0 \text{ on } \mathbf{S}. \tag{4.13}$$

Proof. We organize the proof into two steps.

1. We first show the classical subsolution property:

$$\min\{-(\partial_t+\mathcal{A})V,\ V-g\} \ge 0 \text{ on } \mathbf{S}. \tag{4.14}$$

The inequality $V - g \geq 0$ is obvious as the constant stopping rule $\tau = t \in \mathcal{T}_{[t,T]}$ is admissible. Next, for $(t_0, x_0) \in \mathbf{S}$, consider the stopping times

$$\theta_h := \inf\left\{t > t_0 : \ (t, X_t^{t_0,x_0}) \notin [t_0, t_0 + h] \times B\right\}, h > 0,$$

where B is the unit ball of $\mathbb{R}^n$ centered at x_0. Then $\theta_h \in \mathcal{T}_{[t,T)}^t$ for sufficiently small h, and it follows from (4.10) and the continuity of V that

$$V(t_0, x_0) \geq \mathbb{E}\left[V\left(\theta_h, X_{\theta_h}\right)\right].$$

We next apply Itô's formula, and observe that the expected value of the diffusion term vanishes because (t, X_t) lies in the compact subset $[t_0, t_0 + h] \times B$ for $t \in [t_0, \theta_h]$. Then

$$\mathbb{E}\left[\frac{-1}{h}\int_{t_0}^{\theta_h}(\partial_t + \mathcal{A})V(t, X_t^{t_0,x_0})\mathrm{d}t\right] \geq 0.$$

Clearly, there exists $\hat{h}_\omega > 0$, depending on ω, $\theta_h = h$ for $h \leq \hat{h}_\omega$. Then, it follows from the mean value theorem that the expression inside the expectation converges $\mathbb{P}$–a.s. to $-(\partial_t + \mathcal{A})V(t_0, x_0)$, and we conclude by dominated convergence that $-(\partial_t + \mathcal{A})V(t_0, x_0) \geq 0$.

2. In order to complete the proof, we use a contradiction argument, assuming that

$$V(t_0, x_0) > g(t_0, x_0) \text{ and } -(\partial_t + \mathcal{A})V(t_0, x_0) > 0 \text{ at some } (t_0, x_0) \in \mathbf{S} \quad (4.15)$$

and we work towards a contradiction of (4.9). Introduce the function

$$\varphi(t, x) := V(t, x) + \varepsilon(|x - x_0|^4 + |t - t_0|^2) \text{ for } (t, x) \in \mathbf{S}.$$

Then, it follows from (4.15) that for a sufficiently small $\varepsilon > 0$, we may find $h > 0$ and $\delta > 0$ such that

$$V \geq g + \delta \text{ and } -(\partial_t + \mathcal{A})\varphi \geq 0 \text{ on } \mathcal{N}_h := [t_0, t_0 + h] \times hB. \quad (4.16)$$

Moreover,

$$-\gamma := \max_{\partial\mathcal{N}_h}(V - \varphi) < 0. \quad (4.17)$$

Next, let

$$\theta := \inf\left\{t > t_0 : \ \left(t, X_t^{t_0,x_0}\right) \notin \mathcal{N}_h\right\}.$$

For an arbitrary stopping rule $\tau \in \mathcal{T}_{[t,T]}^t$, we compute by Itô's formula that

$$
\begin{aligned}
\mathbb{E}\left[V\left(\tau\wedge\theta, X_{\tau\wedge\theta}\right) - V(t_0,x_0)\right] &= \mathbb{E}\left[(V-\varphi)\left(\tau\wedge\theta, X_{\tau\wedge\theta}\right)\right] \\
&\quad + \mathbb{E}\left[\varphi\left(\tau\wedge\theta, X_{\tau\wedge\theta}\right) - \varphi(t_0,x_0)\right] \\
&= \mathbb{E}\left[(V-\varphi)\left(\tau\wedge\theta, X_{\tau\wedge\theta}\right)\right] \\
&\quad + \mathbb{E}\left[\int_{t_0}^{\tau\wedge\theta} (\partial_t + \mathcal{A})\varphi(t, X_t^{t_0,x_0})\mathrm{d}t\right],
\end{aligned}
$$

where the diffusion term has zero expectation because the process $(t, X_t^{t_0,x_0})$ is confined to the compact subset $\mathcal{N}_h$ on the stochastic interval $[t_0, \tau\wedge\theta]$. Since $-(\partial_t + \mathcal{A})\varphi \geq 0$ on $\mathcal{N}_h$ by (4.16), this provides

$$
\begin{aligned}
\mathbb{E}\left[V\left(\tau\wedge\theta, X_{\tau\wedge\theta}\right) - V(t_0,x_0)\right] &\leq \mathbb{E}\left[(V-\varphi)\left(\tau\wedge\theta, X_{\tau\wedge\theta}\right)\right] \\
&\leq -\gamma\mathbb{P}[\tau\geq\theta],
\end{aligned}
$$

by (4.17). Then, since $V \geq g + \delta$ on $\mathcal{N}_h$ by (4.16)

$$
\begin{aligned}
V(t_0,x_0) &\geq \gamma\mathbb{P}[\tau\geq\theta] + \mathbb{E}\left[\left(g(X_\tau^{t_0,x_0}) + \delta\right)\mathbf{1}_{\{\tau<\theta\}} + V\left(\theta, X_\theta^{t_0,x_0}\right)\mathbf{1}_{\{\tau\geq\theta\}}\right] \\
&\geq (\gamma\wedge\delta) + \mathbb{E}\left[g(X_\tau^{t_0,x_0})\mathbf{1}_{\{\tau<\theta\}} + V\left(\theta, X_\theta^{t_0,x_0}\right)\mathbf{1}_{\{\tau\geq\theta\}}\right].
\end{aligned}
$$

By the arbitrariness of $\tau \in \mathcal{T}_{[t,T]}^t$ and the continuity of V, this provides the desired contradiction of (4.9). □

Remark 4.6. In the context of the optimal stopping $\bar{V}$ introduced in Remark 4.1, we may derive the dynamic programming equation as the infinitesimal counterpart of the dynamic programming principle of Remark 4.4. Following the same line of argument as in the previous proof, it follows that for $\bar{V} \in C^{1,2}([0,T),\mathbb{R}^n)$, the dynamic programming equation is given by the obstacle problem:

$$
\min\left\{-\partial_t V - \mathcal{L}V - f, V - g\right\} = 0 \text{ on } \mathbf{S}. \tag{4.18}
$$

4.4 Regularity of the Value Function

4.4.1 Finite Horizon Optimal Stopping

In this subsection, we consider the case $T < \infty$. Similar to the continuity result of Proposition 3.7 for the stochastic control framework, the following continuity result is obtained as a consequence of the flow continuity of Theorem 2.4 together with the dynamic programming principle.

Proposition 4.7. *Assume g is Lipschitz continuous, and let $T < \infty$. Then, there is a constant C such that*

$$\left|V(t,x) - V(t',x')\right| \le C\left(|x-x'| + \sqrt{|t-t'|}\right) \text{ for all } (t,x),(t',x') \in \mathbf{S}.$$

Proof. (i) For $t \in [0,T]$ and $x, x' \in \mathbb{R}^n$, it follows from the Lipschitz property of g that

$$\begin{aligned}\left|V(t,x) - V(t,x')\right| &\le \text{Const} \sup_{\tau \in \mathcal{T}_{[t,T]}} \mathbb{E}\left|X_\tau^{t,x} - X_\tau^{t,x'}\right| \\ &\le \text{Const}\, \mathbb{E} \sup_{t \le s \le T} \left|X_\tau^{t,x} - X_\tau^{t,x'}\right| \\ &\le \text{Const}\, |x - x'|\end{aligned}$$

by the flow continuity result of Theorem 2.4.

(ii) To prove the Hölder continuity result in t, we argue as in the proof of Proposition 3.7 using the dynamic programming principle of Theorem 4.3.

(1) We first observe that, whenever the stopping time $\theta = t' > t$ is constant (i.e. deterministic), the dynamic programming principle (4.9) and (4.10) holds true if the semicontinuous envelopes are taken with respect to the variable x, with fixed time variable. Since V is continuous in x by the first part of this proof, we deduce that

$$V(t,x) = \sup_{\tau \in \mathcal{T}_{[t,T]}^t} \mathbb{E}\left[\mathbf{1}_{\{\tau < t'\}} g\left(X_\tau^{t,x}\right) + \mathbf{1}_{\{\tau \ge t'\}} V\left(t', X_{t'}^{t,x}\right)\right]. \tag{4.19}$$

(2) We then estimate that

$$\begin{aligned} 0 &\le V(t,x) - \mathbb{E}\left[V\left(t', X_{t'}^{t,x}\right)\right] \\ &\le \sup_{\tau \in \mathcal{T}_{[t,T]}^t} \mathbb{E}\left[\mathbf{1}_{\{\tau < t'\}}\left(g\left(X_\tau^{t,x}\right) - V\left(t', X_{t'}^{t,x}\right)\right)\right] \\ &\le \sup_{\tau \in \mathcal{T}_{[t,T]}^t} \mathbb{E}\left[\mathbf{1}_{\{\tau < t'\}}\left(g\left(X_\tau^{t,x}\right) - g\left(X_{t'}^{t,x}\right)\right)\right], \end{aligned}$$

where the last inequality follows from the fact that $V \ge g$. Using the Lipschitz property of g, this provides:

$$\begin{aligned} 0 \le V(t,x) - \mathbb{E}\left[V\left(t', X_{t'}^{t,x}\right)\right] &\le \text{Const}\, \mathbb{E}\left[\sup_{t \le s \le t'} \left|X_s^{t,x} - X_{t'}^{t,x}\right|\right] \\ &\le \text{Const}\,(1 + |x|)\sqrt{t'-t} \end{aligned}$$

by the flow continuity result of Theorem 2.4. Using this estimate together with the Lipschitz property proved in (i) above, this provides:

$$\begin{aligned}|V(t,x)-V(t',x)| &\leq \left|V(t,x)-\mathbb{E}\left[V\left(t',X_{t'}^{t,x}\right)\right]\right| \\ &\quad + \left|\mathbb{E}\left[V\left(t',X_{t'}^{t,x}\right)\right]-V(t',x)\right| \\ &\leq \text{Const}\left((1+|x|)\sqrt{t'-t}+\mathbb{E}\left|X_{t'}^{t,x}-x\right|\right) \\ &\leq \text{Const}\,(1+|x|)\sqrt{t'-t},\end{aligned}$$

by using again Theorem 2.4. □

4.4.2 *Infinite Horizon Optimal Stopping*

In this section, the state process X is defined by a homogeneous scalar diffusion:

$$\mathrm{d}X_t = b(X_t)\mathrm{d}t + \sigma(X_t)\mathrm{d}W_t. \tag{4.20}$$

We introduce the hitting times:

$$H_b^x := \inf\left\{t > 0 :\ X^{0,x} = b\right\},$$

and we assume that the process X is regular, i.e.,

$$\mathbb{P}\left[H_b^x < \infty\right] > 0 \quad \text{for all } x, b \in \mathbb{R}, \tag{4.21}$$

which means that there is no subinterval of $\mathbb{R}$ from which the process X cannot exit.

We consider the infinite horizon optimal stopping problem:

$$V(x) := \sup_{\tau\in\mathcal{T}} \mathbb{E}\left[\mathrm{e}^{-\beta\tau} g\left(X_\tau^{0,x}\right)\mathbf{1}_{\{\tau<\infty\}}\right], \tag{4.22}$$

where $\mathcal{T} := \mathcal{T}_{[0,\infty]}$, and $\beta > 0$ is the discount rate parameter.

According to Theorem 4.3, the dynamic programming equation corresponding to this optimal stopping problem is the obstacle problem:

$$\min\left\{\beta v - \mathcal{A}v, v - g\right\} = 0,$$

where the differential operator in the present homogeneous context is given by the generator of the diffusion:

$$\mathcal{A}v := bv' + \frac{1}{2}\sigma^2 v''. \tag{4.23}$$

The ordinary differential equation

$$\mathcal{A}v - \beta v = 0 \tag{4.24}$$

has two positive linearly independent solutions

$$\psi, \phi \geq 0 \text{ such that } \psi, \textit{ strictly increasing } \phi \text{ strictly decreasing.} \tag{4.25}$$

Clearly ψ and ϕ are uniquely determined up to a positive constant, and all other solution of (4.24) can be expressed as a linear combination of ψ and ϕ.

The following result follows from an immediate application of Itô's formula.

Lemma 4.8. *For any $b_1 < b_2$, we have*

$$\mathbb{E}\left[e^{-\beta H_{b_1}^x} \mathbf{1}_{\{H_{b_1}^x \leq H_{b_2}^x\}}\right] = \frac{\psi(x)\phi(b_2) - \psi(b_2)\phi(x)}{\psi(b_1)\phi(b_2) - \psi(b_2)\phi(b_1)},$$

$$\mathbb{E}\left[e^{-\beta H_{b_2}^x} \mathbf{1}_{\{H_{b_1}^x \geq H_{b_2}^x\}}\right] = \frac{\psi(b_1)\phi(x) - \psi(x)\phi(b_1)}{\psi(b_1)\phi(b_2) - \psi(b_2)\phi(b_1)}.$$

We now show that the value function V is concave up to some change of variable and provide conditions under which V is C^1 across the exercise boundary, i.e., the boundary between the exercise and the continuation regions.

For the next result, we observe that the function (ψ/ϕ) is continuous and strictly increasing by (4.25), and therefore invertible. To prepare for its statement, let us observe that, on any compact subset of the continuation region, we have $V = A\psi + B\psi$ for some constants $A, B \in \mathbb{R}$. Then, $(V/\phi) \circ (\psi/\phi)^{-1}(x) = Ax + b$ is affine on any compact subset of the continuation region. The following result examines the nature of $(V/\phi) \circ (\psi/\phi)^{-1}$ on the entire domain.

Theorem 4.9. (i) *The function $(V/\phi) \circ (\psi/\phi)^{-1}$ is concave. In particular, V is continuous on $\mathbb{R}$.*

(ii) *Let x_0 be such that $V(x_0) = g(x_0)$, and assume that g, ψ and ϕ are differentiable at x_0. Then V is differentiable at x_0, and $V'(x_0) = g'(x_0)$.*

Proof. Denote $F := \psi/\phi$. For (i), it is sufficient to prove that:

$$\frac{\frac{V}{\phi}(x) - \frac{V}{\phi}(b_1)}{F(x) - F(b_1)} \leq \frac{\frac{V}{\phi}(b_2) - \frac{V}{\phi}(x)}{F(b_2) - F(x)} \quad \text{for all } b_1 < x < b_2. \tag{4.26}$$

For $\varepsilon > 0$, consider the ε–optimal stopping rules $\tau_1, \tau_2 \in \mathcal{T}$ for the problems $V(b_1)$ and $V(b_2)$:

$$\mathbb{E}\left[e^{-\beta\tau_i} g\left(X_{\tau_i}^{0,x}\right)\right] \geq V(b_i) - \varepsilon \quad \text{for } i = 1, 2.$$

We next define the stopping time

$$\tau^\varepsilon := \left(H_{b_1}^x + \tau_1 \circ \theta_{H_{b_1}^x}\right) \mathbf{1}_{\{H_{b_1}^x < H_{b_2}^x\}} + \left(H_{b_2}^x + \tau_2 \circ \theta_{H_{b_2}^x}\right) \mathbf{1}_{\{H_{b_2}^x < H_{b_1}^x\}},$$

where θ denotes the shift operator on the canonical space. In words, the stopping rule τ^ε uses the ε−optimal stopping rule τ_1 if the level b_1 is reached before the level b_2 and the ε−optimal stopping rule τ_2 otherwise. Then, it follows from the strong Markovian property that

$$\begin{aligned} V(x) &\geq \mathbb{E}\left[\mathrm{e}^{-\beta\tau^\varepsilon} g\left(X^{0,x}_{\tau^\varepsilon}\right)\right] \\ &= \mathbb{E}\left[\mathrm{e}^{-\beta H^x_{b_1}} \mathbb{E}\left[\mathrm{e}^{-\beta\tau_1} g\left(X^{0,b_1}_{\tau_1}\right)\right] \mathbf{1}_{\{H^x_{b_1}<H^x_{b_2}\}}\right] \\ &\quad + \mathbb{E}\left[\mathrm{e}^{-\beta H^x_{b_2 1}} \mathbb{E}\left[\mathrm{e}^{-\beta\tau_2} g\left(X^{0,b_2}_{\tau_2}\right)\right] \mathbf{1}_{\{H^x_{b_2}<H^x_{b_1}\}}\right] \\ &\geq \left(V(b_1)-\varepsilon\right) \mathbb{E}\left[\mathrm{e}^{-\beta H^x_{b_1}} \mathbf{1}_{\{H^x_{b_1}<H^x_{b_2}\}}\right] \\ &\quad + \left(V(b_2)-\varepsilon\right) \mathbb{E}\left[\mathrm{e}^{-\beta H^x_{b_2}} \mathbf{1}_{\{H^x_{b_2}<H^x_{b_1}\}}\right]. \end{aligned}$$

Sending $\varepsilon \searrow 0$, this provides

$$V(x) \geq V(b_1)\mathbb{E}\left[\mathrm{e}^{-\beta H^x_{b_1}} \mathbf{1}_{\{H^x_{b_1}<H^x_{b_2}\}}\right] + V(b_2)\mathbb{E}\left[\mathrm{e}^{-\beta H^x_{b_2}} \mathbf{1}_{\{H^x_{b_2}<H^x_{b_1}\}}\right].$$

By using the explicit expressions of Lemma 4.8 above, this provides

$$\frac{V(x)}{\phi(x)} \geq \frac{V(b_1)}{\phi(b_1)} \frac{F(b_2)-F(x)}{F(b_2)-F(b_1)} + \frac{V(b_2)}{\phi(b_2)} \frac{F(x)-F(b_1)}{F(b_2)-F(b_1)},$$

which implies (4.26).

(ii) We next prove the smoothfit result. Let x_0 be such that $V(x_0) = g(x_0)$. Then, since $V \geq g$, ψ is strictly increasing and $\phi \geq 0$ is strictly decreasing, it follows from (4.26) that:

$$\begin{aligned} \frac{\frac{g}{\phi}(x_0+\varepsilon)-\frac{g}{\phi}(x_0)}{\frac{\psi}{\phi}(x_0+\varepsilon)-\frac{\psi}{\phi}(x_0)} &\leq \frac{\frac{V}{\phi}(x_0+\varepsilon)-\frac{V}{\phi}(x_0)}{\frac{\psi}{\phi}(x_0+\varepsilon)-\frac{\psi}{\phi}(x_0)} \\ &\leq \frac{\frac{V}{\phi}(x_0-\delta)-\frac{V}{\phi}(x_0)}{\frac{\psi}{\phi}(x_0-\delta)-\frac{\psi}{\phi}(x_0)} \leq \frac{\frac{g}{\phi}(x_0-\delta)-\frac{g}{\phi}(x_0)}{\frac{\psi}{\phi}(x_0-\delta)-\frac{\psi}{\phi}(x_0)}, \end{aligned} \tag{4.27}$$

for all $\varepsilon > 0$, $\delta > 0$. Multiplying by $((\psi/\phi)(x_0+\varepsilon)-(\psi/\phi)(x_0))/\varepsilon$, this implies that:

$$\begin{aligned} \frac{\frac{g}{\phi}(x_0+\varepsilon)-\frac{g}{\phi}(x_0)}{\varepsilon} &\leq \frac{\frac{V}{\phi}(x_0+\varepsilon)-\frac{V}{\phi}(x_0)}{\varepsilon} \\ &\leq \frac{\Delta^+(\varepsilon)}{\Delta^-(\delta)} \frac{\frac{g}{\phi}(x_0-\delta)-\frac{g}{\phi}(x_0)}{\delta}, \end{aligned} \tag{4.28}$$

where

$$\Delta^+(\varepsilon) := \frac{\frac{\psi}{\phi}(x_0+\varepsilon) - \frac{\psi}{\phi}(x_0)}{\varepsilon} \text{ and } \Delta^-(\delta) := \frac{\frac{\psi}{\phi}(x_0-\delta) - \frac{\psi}{\phi}(x_0)}{\delta}.$$

We next consider two cases:

- If $(\psi/\phi)'(x_0) \neq 0$, then we may take $\varepsilon = \delta$ and send $\varepsilon \searrow 0$ in (4.28) to obtain:

$$\frac{d^+(\frac{V}{\phi})}{dx}(x_0) = \left(\frac{g}{\phi}\right)'(x_0). \tag{4.29}$$

- If $(\psi/\phi)'(x_0) = 0$, then, we use the fact that for every sequence $\varepsilon_n \searrow 0$, there is a subsequence $\varepsilon_{n_k} \searrow 0$ and $\delta_k \searrow 0$ such that $\Delta^+(\varepsilon_{n_k}) = \Delta^-(\delta_k)$. Then (4.28) reduces to:

$$\frac{\frac{g}{\phi}(x_0+\varepsilon_{n_k}) - \frac{g}{\phi}(x_0)}{\varepsilon_{n_k}} \leq \frac{\frac{V}{\phi}(x_0+\varepsilon_{n_k}) - \frac{V}{\phi}(x_0)}{\varepsilon_{n_k}} \leq \frac{\frac{g}{\phi}(x_0-\delta_k) - \frac{g}{\phi}(x_0)}{\delta_k},$$

  and therefore

$$\frac{\frac{V}{\phi}(x_0+\varepsilon_{n_k}) - \frac{V}{\phi}(x_0)}{\varepsilon_{n_k}} \longrightarrow \left(\frac{g}{\phi}\right)'(x_0).$$

  By the arbitrariness of the sequence $(\varepsilon_n)_n$, this provides (4.29).

Similarly, multiplying (4.27) by $((\psi/\phi)(x_0) - (\psi/\phi)(x_0-\delta))/\delta$, and arguing as above, we obtain:

$$\frac{d^-(\frac{V}{\phi})}{dx}(x_0) = \left(\frac{g}{\phi}\right)'(x_0),$$

thus completing the proof. □

4.4.3 An Optimal Stopping Problem with Nonsmooth Value

We consider the example

$$X_s^{t,x} := x + (W_t - W_s) \quad \text{for } s \geq t.$$

Let $g : \mathbb{R} \longrightarrow \mathbb{R}_+$ be a measurable nonnegative function and consider the infinite horizon optimal stopping problem:

$$\begin{aligned} V(t,x) &:= \sup_{\tau \in \mathcal{T}_{[t,\infty]}} \mathbb{E}\left[g\left(X_\tau^{t,x}\right)\mathbf{1}_{\{\tau<\infty\}}\right] \\ &= \sup_{\tau \in \mathcal{T}_{[t,\infty)}} \mathbb{E}\left[g\left(X_\tau^{t,x}\right)\right], \end{aligned}$$

where the last equality follows from the nonnegativity of g together with Fatou's Lemma. Let us assume that $V \in C^{1,2}(\mathbf{S})$ and work towards a contradiction. We first observe by the homogeneity of the problem that $V(t,x) = V(x)$ is independent of t. Moreover, it follows from Theorem 4.5 that V is concave in x and $V \geq g$. Then

$$V \geq g^{\mathrm{conc}}, \tag{4.30}$$

where g^{conc} is the concave envelope of g. If $g^{\mathrm{conc}} = \infty$, then $V = \infty$. We then continue in the more interesting case where $g^{\mathrm{conc}} < \infty$.

By the Jensen inequality and the non-negativity of g, the process $\left\{g\left(X_s^{t,x}\right), s \geq t\right\}$ is a supermartingale, and

$$V(t,x) \leq \sup_{\tau \in \mathcal{T}_{[t,T]}} \mathbb{E}\left[g^{\mathrm{conc}}\left(X_\tau^{t,x}\right)\right] \leq g^{\mathrm{conc}}(x).$$

Hence, $V = g^{\mathrm{conc}}$, and we obtain the required contradiction whenever g^{conc} is not differentiable at some point of $\mathbb{R}$.

Chapter 5
Solving Control Problems by Verification

In this chapter, we present a general argument, based on Itô's formula, which allows to show that some "guess" of the value function is indeed equal to the unknown value function. Namely, given a smooth solution v of the dynamic programming equation, we give sufficient conditions which allow to conclude that v coincides with the value function V. This is the so-called *verification argument*. The statement of this result is heavy, but its proof is simple and relies essentially on Itô's formula. However, depending on the problem in hand, the verification of the conditions which must be satisfied by the candidate solution can be difficult.

The verification argument will be provided in the contexts of stochastic control and optimal stopping problems. We conclude this chapter with some examples.

5.1 The Verification Argument for Stochastic Control Problems

We recall the stochastic control problem formulation of Sect. 3.1. The set of admissible control processes $\mathcal{U}_0 \subset \mathcal{U}$ is the collection of all progressively measurable processes with values in the subset $U \subset \mathbb{R}^k$. For every admissible control process $\nu \in \mathcal{U}_0$, the controlled process is defined by the stochastic differential equation:

$$\mathrm{d}X_t^\nu = b(t, X_t^\nu, \nu_t)\mathrm{d}t + \sigma(t, X_t^\nu, \nu_t)\mathrm{d}W_t.$$

The gain criterion is given by

$$J(t,x,\nu) := \mathbb{E}\left[\int_t^T \beta^\nu(t,s) f(s, X_s^{t,x,\nu}, \nu_s)\mathrm{d}s + \beta^\nu(t,T) g(X_T^{t,x,\nu})\right],$$

with

$$\beta^\nu(t,s) := \mathrm{e}^{-\int_t^s k(r, X_r^{t,x,\nu}, \nu_r)\mathrm{d}r}.$$

N. Touzi, *Optimal Stochastic Control, Stochastic Target Problems, and Backward SDE*, Fields Institute Monographs 29, DOI 10.1007/978-1-4614-4286-8_5,

The stochastic control problem is defined by the value function:

$$V(t,x) := \sup_{\nu \in \mathcal{U}_0} J(t,x,\nu), \qquad \text{for } (t,x) \in \mathbf{S}. \tag{5.1}$$

We follow the notations of Sect. 3.3. We recall the Hamiltonian $H : \mathbf{S} \times \mathbb{R} \times \mathbb{R}^d \times \mathcal{S}_d$ defined by:

$$\begin{aligned} &H(t,x,r,p,\gamma) \\ &:= \sup_{u \in U} \left\{ -k(t,x,u)r + b(t,x,u)\cdot p + \frac{1}{2}\mathrm{Tr}[\sigma\sigma^{\mathrm{T}}(t,x,u)\gamma] + f(t,x,u) \right\}, \end{aligned}$$

where b and σ satisfy the conditions (3.1)–(3.2), and the coefficients f and k are measurable. From the results of the previous section, the dynamic programming equation corresponding to the stochastic control problem (5.1) is

$$-\partial_t v - H(.,v,Dv,D^2v) = 0 \text{ and } v(T,.) = g. \tag{5.2}$$

A function v will be called a *supersolution* (resp. *subsolution*) of the (5.2) if

$$-\partial_t v - H(.,v,Dv,D^2v) \geq \text{ (resp. } \leq) \ 0 \text{ and } v(T,.) \geq \text{ (resp. } \leq) \ g.$$

The proof of the subsequent result will make use of the following linear second order operator:

$$\begin{aligned} \mathcal{L}^u\varphi(t,x) &:= -k(t,x,u)\varphi(t,x) + b(t,x,u)\cdot D\varphi(t,x) \\ &\quad + \frac{1}{2}\mathrm{Tr}\left[\sigma\sigma^{\mathrm{T}}(t,x,u)D^2\varphi(t,x)\right], \end{aligned}$$

which corresponds to the process $\{\beta^u(0,t)X_t^u,\ t \geq 0\}$ controlled by the constant control process u, in the sense that

$$\begin{aligned} \beta^\nu(0,s)\varphi(s,X_s^\nu) - \beta^\nu(0,t)\varphi(t,X_t^\nu) &= \int_t^s \beta^\nu(0,r)\left(\partial_t + \mathcal{L}^{\nu_r}\right)\varphi(r,X_r^\nu)\mathrm{d}r \\ &\quad + \int_t^s \beta^\nu(0,r)D\varphi(r,X_r^\nu)\cdot\sigma(r,X_r^\nu,\nu_r)\mathrm{d}W_r, \end{aligned}$$

for every $t \leq s$ and smooth function $\varphi \in C^{1,2}([t,s],\mathbb{R}^d)$ and each admissible control process $\nu \in \mathcal{U}_0$. The last expression is an immediate application of Itô's formula.

Theorem 5.1. *Let $T < \infty$, and $v \in C^{1,2}([0,T),\mathbb{R}^d) \cap C([0,T]\times\mathbb{R}^d)$. Assume that $\|k^-\|_\infty < \infty$ and v and f have quadratic growth, i.e., there is a constant C such that*

$$|f(t,x,u)| + |v(t,x)| \ \leq \ C(1+|x|^2+|u|),\ (t,x,u) \in [0,T)\times\mathbb{R}^d\times U.$$

(i) *Suppose that v is a supersolution of* (5.2). *Then $v \geq V$ on $[0,T] \times \mathbb{R}^d$.*
(ii) *Let v be a solution of* (5.2), *and assume that there exists a minimizer $\hat{u}(t,x)$ of $u \longmapsto \mathcal{L}^u v(t,x) + f(t,x,u)$ such that*

- $0 = \partial_t v(t,x) + \mathcal{L}^{\hat{u}(t,x)} v(t,x) + f\big(t,x,\hat{u}(t,x)\big)$
- *The stochastic differential equation*

$$dX_s = b\left(s, X_s, \hat{u}(s,X_s)\right) ds + \sigma\left(s, X_s, \hat{u}(s,X_s)\right) dW_s$$

defines a unique solution X for each given initial data $X_t = x$
- *The process $\hat{\nu}_s := \hat{u}(s,X_s)$ is a well-defined control process in $\mathcal{U}_0$.*

Then $v = V$, and $\hat{\nu}$ is an optimal Markovian control process.

Proof. Let $\nu \in \mathcal{U}_0$ be an arbitrary control process, X the associated state process with initial date $X_t = x$, and define the stopping time

$$\theta_n := (T - n^{-1}) \wedge \inf\{s > t \ : \ |X_s - x| \geq n\}.$$

By Itô's formula, we have

$$v(t,x) = \beta(t,\theta_n) v\left(\theta_n, X_{\theta_n}\right) - \int_t^{\theta_n} \beta(t,r)(\partial_t + \mathcal{L}^{\nu_r}) v(r,X_r) dr$$
$$- \int_t^{\theta_n} \beta(t,r) Dv(r,X_r) \cdot \sigma(r,X_r,\nu_r) dW_r.$$

Observe that $(\partial_t + \mathcal{L}^{\nu_r})v + f(\cdot,\cdot,u) \leq \partial_t v + H(\cdot,\cdot,v,Dv,D^2v) \leq 0$, and that the integrand in the stochastic integral is bounded on $[t,\theta_n]$, a consequence of the continuity of Dv, σ, and the condition $\|k^-\|_\infty < \infty$. Then:

$$v(t,x) \geq \mathbb{E}\left[\beta(t,\theta_n) v\left(\theta_n, X_{\theta_n}\right) + \int_t^{\theta_n} \beta(t,r) f(r,X_r,\nu_r) dr\right]. \tag{5.3}$$

We now take the limit as n increases to infinity. Since $\theta_n \longrightarrow T$ a.s. and

$$\left|\beta(t,\theta_n) v\left(\theta_n, X_{\theta_n}\right) + \int_t^{\theta_n} \beta(t,r) f(r,X_r,\nu_r) dr\right|$$
$$\leq C e^{T\|k^-\|_\infty}\left(1 + |X_{\theta_n}|^2 + T + \int_t^T |X_s|^2 ds\right)$$
$$\leq C e^{T\|k^-\|_\infty}(1+T)\left(1 + \sup_{t \leq s \leq T} |X_s|^2 + \int_t^T |\nu_s|^2 ds\right) \in \mathbb{L}^1,$$

by the estimate (3.5) of Theorem 3.1, it follows from the dominated convergence that

$$\begin{aligned} v(t,x) &\geq \mathbb{E}\left[\beta(t,T)v(T,X_T) + \int_t^T \beta(t,r)f(r,X_r,\nu_r)\mathrm{d}r\right] \\ &\geq \mathbb{E}\left[\beta(t,T)g(X_T) + \int_t^T \beta(t,r)f(r,X_r,\nu_r)\mathrm{d}r\right], \end{aligned}$$

where the last inequality uses the condition $v(T,\cdot) \geq g$. Since the control $\nu \in \mathcal{U}_0$ is arbitrary, this completes the proof of (i).

Statement (ii) is proved by repeating the above argument and observing that the control $\hat{\nu}$ achieves equality at the crucial step (5.3). □

Remark 5.2. When U is reduced to a singleton, the optimization problem V is degenerate. In this case, the DPE is linear, and the verification theorem reduces to the so-called *Feynman-Kac formula*.

Notice that the verification theorem assumes the existence of such a solution, and is by no means an existence result. However, it provides uniqueness in the class of functions with quadratic growth.

We now state without proof an existence result for the DPE together with the terminal condition $V(T,\cdot) = g$ (see [23] and the references therein). The main assumption is the so-called *uniform parabolicity* condition:

$$\text{there is a constant } c > 0 \text{ such that}$$
$$\xi' \,\sigma\sigma'(t,x,u)\, \xi \;\geq\; c|\xi|^2 \quad \text{for all } (t,x,u) \in [0,T] \times \mathbb{R}^n \times U. \tag{5.4}$$

In the following statement, we denote by $C_b^k(\mathbb{R}^n)$ the space of bounded functions whose partial derivatives of orders $\leq k$ exist and are bounded continuous. We similarly denote by $C_b^{p,k}([0,T],\mathbb{R}^n)$ the space of bounded functions whose partial derivatives with respect to t, of orders $\leq p$, and with respect to x, of order $\leq k$, exist and are bounded continuously

Theorem 5.3. *Let Condition 5.4 hold, and assume further that:*

- *U is compact*
- *b, σ, and f are in $C_b^{1,2}([0,T],\mathbb{R}^n)$*
- *$g \in C_b^3(\mathbb{R}^n)$*

Then the DPE (3.21) *with the terminal data $V(T,\cdot) = g$ has a unique solution $V \in C_b^{1,2}([0,T] \times \mathbb{R}^n)$.*

5.2 Examples of Control Problems with Explicit Solutions

5.2.1 *Optimal Portfolio Allocation*

We now apply the verification theorem to a classical example in finance, which was introduced by Merton [29, 30], and generated a huge literature since then.

Consider a financial market consisting of a nonrisky asset S^0 and a risky one S. The dynamics of the price processes is given by

$$\mathrm{d}S_t^0 = S_t^0 r\mathrm{d}t \text{ and } \mathrm{d}S_t = S_t[\mu \mathrm{d}t + \sigma \mathrm{d}W_t].$$

Here, r, μ, and σ are some given positive constants, and W is a one-dimensional Brownian motion.

The investment policy is defined by an $\mathbb{F}-$adapted process $\pi = \{\pi_t, t \in [0,T]\}$, where π_t represents the amount invested in the risky asset at time t; The remaining wealth $(X_t - \pi_t)$ is invested in the risky asset. Therefore, the liquidation value of a self-financing strategy satisfies

$$\begin{aligned} \mathrm{d}X_t^\pi &= \pi_t \frac{\mathrm{d}S_t}{S_t} + (X_t^\pi - \pi_t)\frac{\mathrm{d}S_t^0}{S_t^0} \\ &= (rX_t + (\mu - r)\pi_t)\,\mathrm{d}t + \sigma\pi_t \mathrm{d}W_t. \end{aligned} \tag{5.5}$$

Such a process π is said to be admissible if it lies in $\mathcal{U}_0 = \mathbb{H}^2$ which will be referred to as the set of all admissible portfolios. Observe that, in view of the particular form of our controlled process X, this definition agrees with (3.4).

Let γ be an arbitrary parameter in $(0,1)$ and define the *power utility function*:

$$U(x) := x^\gamma \text{ for } x \geq 0.$$

The parameter γ is called the relative risk aversion coefficient.

The objective of the investor is to choose an allocation of his wealth so as to maximize the expected utility of his terminal wealth, i.e.,

$$V(t,x) := \sup_{\pi \in \mathcal{U}_0} \mathbb{E}\left[U(X_T^{t,x,\pi})\right],$$

where $X^{t,x,\pi}$ is the solution of (5.5) with initial condition $X_t^{t,x,\pi} = x$.

The dynamic programming equation corresponding to this problem is

$$\frac{\partial w}{\partial t}(t,x) + \sup_{u \in \mathbb{R}} \mathcal{A}^u w(t,x) = 0, \tag{5.6}$$

where $\mathcal{A}^u$ is the second-order linear operator

$$\mathcal{A}^u w(t,x) := (rx + (\mu - r)u)\,\frac{\partial w}{\partial x}(t,x) + \frac{1}{2}\sigma^2 u^2\,\frac{\partial^2 w}{\partial x^2}(t,x).$$

We next search for a solution of the dynamic programming equation of the form $v(t,x) = x^\gamma h(t)$. Plugging this form of solution into the PDE (5.6), we get the following ordinary differential equation on h:

$$0 = h' + \gamma h \sup_{u\in\mathbb{R}} \left\{ r + (\mu - r)\frac{u}{x} + \frac{1}{2}(\gamma - 1)\sigma^2 \frac{u^2}{x^2} \right\} \tag{5.7}$$

$$= h' + \gamma h \sup_{\delta\in\mathbb{R}} \left\{ r + (\mu - r)\delta + \frac{1}{2}(\gamma - 1)\sigma^2\delta^2 \right\} \tag{5.8}$$

$$= h' + \gamma h \left[r + \frac{1}{2}\frac{(\mu - r)^2}{(1-\gamma)\sigma^2} \right], \tag{5.9}$$

where the maximizer is

$$\hat{u} := \frac{\mu - r}{(1-\gamma)\sigma^2}\; x.$$

Since $v(T,\cdot) = U(x)$, we seek for a function h satisfying the above ordinary differential equation together with the boundary condition $h(T) = 1$. This induces the unique candidate:

$$h(t) := e^{a(T-t)} \text{ with } a := \gamma \left[r + \frac{1}{2}\frac{(\mu - r)^2}{(1-\gamma)\sigma^2} \right].$$

Hence, the function $(t,x) \longmapsto x^\gamma h(t)$ is a classical solution of the HJB equation (5.6). It is easily checked that the conditions of Theorem 5.1 are all satisfied in this context. Then $V(t,x) = x^\gamma h(t)$, and the optimal portfolio allocation policy is given by the linear control process:

$$\hat{\pi}_t = \frac{\mu - r}{(1-\gamma)\sigma^2}\; X_t^{\hat{\pi}}.$$

5.2.2 *Law of Iterated Logarithm for Double Stochastic Integrals*

The main object of this paragraph is Theorem 5.5, reported from [11], which describes the local behavior of double stochastic integrals near the starting point zero. This result will be needed in the problem of hedging under gamma constraints which will be discussed later in these notes. An interesting feature of the proof of

Theorem 5.5 is that it relies on a verification argument. However, the problem does not fit exactly in the setting of Theorem 5.1. Therefore, this is an interesting exercise on the verification concept.

Given a bounded predictable process b, we define the processes

$$Y_t^b := Y_0 + \int_0^t b_r \mathrm{d}W_r \text{ and } Z_t^b := Z_0 + \int_0^t Y_r^b \mathrm{d}W_r, \quad t \geq 0,$$

where Y_0 and Z_0 are some given initial data in $\mathbb{R}$.

Lemma 5.4. *Let λ and T be two positive parameters with $2\lambda T < 1$. Then:*

$$E\left[\mathrm{e}^{2\lambda Z_T^b}\right] \leq E\left[\mathrm{e}^{2\lambda Z_T^1}\right] \text{ for each predictable process } b \text{ with } \|b\|_\infty \leq 1.$$

Proof. We split the argument into three steps

1. We first directly compute that

$$E\left[\mathrm{e}^{2\lambda Z_T^1}\,\middle|\,\mathcal{F}_t\right] = v(t, Y_t^1, Z_t^1),$$

where, for $t \in [0, T]$, and $y, z \in \mathbb{R}$, the function v is given by:

$$\begin{aligned} v(t,y,z) &:= E\left[\exp\left(2\lambda\left\{z + \int_t^T (y + W_u - W_t)\,\mathrm{d}W_u\right\}\right)\right] \\ &= \mathrm{e}^{2\lambda z} E\left[\exp\left(\lambda\left\{2yW_{T-t} + W_{T-t}^2 - (T-t)\right\}\right)\right] \\ &= \mu \exp\left[2\lambda z - \lambda(T-t) + 2\mu^2\lambda^2(T-t)y^2\right], \end{aligned}$$

where $\mu := [1 - 2\lambda(T-t)]^{-1/2}$. Observe that

$$\text{the function } v \text{ is strictly convex in } y, \tag{5.10}$$

and

$$yD_{yz}^2 v(t,y,z) = 8\mu^2\lambda^3(T-t)\, v(t,y,z)\, y^2 \geq 0. \tag{5.11}$$

2. For an arbitrary real parameter β, we denote by $\mathcal{A}^\beta$ the generator the process (Y^b, Z^b):

$$\mathcal{A}^\beta := \frac{1}{2}\beta^2 D_{yy}^2 + \frac{1}{2}y^2 D_{zz}^2 + \beta y D_{yz}^2.$$

In this step, we intend to prove that for all $t \in [0, T]$ and $y, z \in \mathbb{R}$,

$$\left[\partial_t v + \max_{|\beta| \leq 1} \mathcal{A}^\beta v\right](t,y,z) = \left[\partial_t v + \mathcal{A}^1 v(t,y,z)\right] = 0. \tag{5.12}$$

The second equality follows from the fact that $\{v(t, Y_t^1, Z_t^1),\ t \leq T\}$ is a martingale . As for the first equality, we see from (5.10) and (5.11) that 1 is a maximizer of both functions $\beta \longmapsto \beta^2 D_{yy}^2 v(t, y, z)$ and $\beta \longmapsto \beta y D_{yz}^2 v(t, y, z)$ on $[-1, 1]$.

3. Let b be some given predictable process valued in $[-1, 1]$, and define the sequence of stopping times

$$\tau_k := T \wedge \inf\{t \geq 0 : |Y_t^b| + |Z_t^b| \geq k\}, \quad k \in \mathbb{N}.$$

By Itô's lemma and (5.12), it follows that

$$\begin{aligned} v(0, Y_0, Z_0) &= v\left(\tau_k, Y_{\tau_k}^b, Z_{\tau_k}^b\right) - \int_0^{\tau_k} [bD_y v + yD_z v]\left(t, Y_t^b, Z_t^b\right) \mathrm{d}W_t \\ &\quad - \int_0^{\tau_k} (\partial_t + \mathcal{A}^{b_t}) v\left(t, Y_t^b, Z_t^b\right) \mathrm{d}t \\ &\geq v\left(\tau_k, Y_{\tau_k}^b, Z_{\tau_k}^b\right) - \int_0^{\tau_k} [bD_y v + yD_z v]\left(t, Y_t^b, Z_t^b\right) \mathrm{d}W_t. \end{aligned}$$

Taking expected values and sending k to infinity, we get by Fatou's lemma

$$\begin{aligned} v(0, Y_0, Z_0) &\geq \liminf_{k \to \infty} E\left[v\left(\tau_k, Y_{\tau_k}^b, Z_{\tau_k}^b\right)\right] \\ &\geq E\left[v\left(T, Y_T^b, Z_T^b\right)\right] = E\left[e^{2\lambda Z_T^b}\right], \end{aligned}$$

which proves the lemma. □

We are now able to prove the law of the iterated logarithm for double stochastic integrals by a direct adaptation of the case of the Brownian motion. Set

$$h(t) := 2t \log\log \frac{1}{t} \quad \text{for } t > 0 .$$

Theorem 5.5. *Let b be a predictable process valued in a bounded interval $[\beta_0, \beta_1]$ for some real parameters $0 \leq \beta_0 < \beta_1$, and $X_t^b := \int_0^t \int_0^u b_v \mathrm{d}W_v \mathrm{d}W_u$. Then*

$$\beta_0 \leq \limsup_{t \searrow 0} \frac{2X_t^b}{h(t)} \leq \beta_1 \quad a.s.$$

Proof. We first show that the first inequality is an easy consequence of the second one. Set $\bar{\beta} := (\beta_0 + \beta_1)/2 \geq 0$, and set $\delta := (\beta_1 - \beta_0)/2$. By the law of the iterated logarithm for the Brownian motion, we have

$$\bar{\beta} = \limsup_{t \searrow 0} \frac{2X_t^{\bar{\beta}}}{h(t)} \leq \delta \limsup_{t \searrow 0} \frac{2X_t^{\tilde{b}}}{h(t)} + \limsup_{t \searrow 0} \frac{2X_t^b}{h(t)},$$

where $\tilde{b} := \delta^{-1}(\bar{\beta}-b)$ is valued in $[-1,1]$. It then follows from the second inequality that

$$\limsup_{t\searrow 0} \frac{2X_t^b}{h(t)} \geq \bar{\beta} - \delta \;=\; \beta_0.$$

We now prove the second inequality. Clearly, we can assume with no loss of generality that $\|b\|_\infty \leq 1$. Let $T > 0$ and $\lambda > 0$ be such that $2\lambda T < 1$. It follows from Doob's maximal inequality for submartingales that for all $\alpha \geq 0$,

$$\begin{aligned} P\left[\max_{0\leq t\leq T} 2X_t^b \;\geq\; \alpha\right] &= P\left[\max_{0\leq t\leq T} \exp(2\lambda X_t^b) \;\geq\; \exp(\lambda\alpha)\right] \\ &\leq \mathrm{e}^{-\lambda\alpha} E\left[\mathrm{e}^{2\lambda X_T^b}\right]. \end{aligned}$$

In view of Lemma 5.4, this provides

$$\begin{aligned} P\left[\max_{0\leq t\leq T} 2X_t^b \;\geq\; \alpha\right] &\leq \mathrm{e}^{-\lambda\alpha} E\left[\mathrm{e}^{2\lambda X_T^1}\right] \\ &= \mathrm{e}^{-\lambda(\alpha+T)}(1-2\lambda T)^{-\frac{1}{2}}. \end{aligned} \tag{5.13}$$

We have then reduced the problem to the case of the Brownian motion, and the rest of this proof is identical to the first half of the proof of the law of the iterated logarithm for the Brownian motion. Take $\theta, \eta \in (0,1)$ and set for all $k \in \mathbb{N}$

$$\alpha_k := (1+\eta)^2 h(\theta^k) \text{ and } \lambda_k := [2\theta^k(1+\eta)]^{-1}.$$

Applying (5.13), we see that for all $k \in \mathbb{N}$,

$$P\left[\max_{0\leq t\leq \theta^k} 2X_t^b \;\geq\; (1+\eta)^2 h(\theta^k)\right] \;\leq\; \mathrm{e}^{-1/2(1+\eta)}\left(1+\eta^{-1}\right)^{\frac{1}{2}}(-k\log\theta)^{-(1+\eta)}.$$

Since $\sum_{k\geq 0} k^{-(1+\eta)} < \infty$, it follows from the Borel-Cantelli lemma that, for almost all $\omega \in \Omega$, there exists a natural number $K^{\theta,\eta}(\omega)$ such that for all $k \geq K^{\theta,\eta}(\omega)$,

$$\max_{0\leq t\leq \theta^k} 2X_t^b(\omega) < (1+\eta)^2 h(\theta^k).$$

In particular, for all $t \in (\theta^{k+1}, \theta^k]$,

$$2X_t^b(\omega) < (1+\eta)^2 h(\theta^k) \;\leq\; (1+\eta)^2 \frac{h(t)}{\theta}.$$

Hence,

$$\limsup_{t\searrow 0} \frac{2X_t^b}{h(t)} < \frac{(1+\eta)^2}{\theta} \quad \text{a.s.},$$

and the required result follows by letting θ tend to 1 and η to 0 along the rationals. □

5.3 The Verification Argument for Optimal Stopping Problems

In this section, we develop the verification argument for finite horizon optimal stopping problems. Let $T > 0$ be a finite time horizon and $X^{t,x}$ denote the solution of the stochastic differential equation:

$$X_s^{t,x} = x + \int_t^s b(s, X_s^{t,x})\mathrm{d}s + \int_t^s \sigma(s, X_s^{t,x})\mathrm{d}W_s, \tag{5.14}$$

where b and σ satisfy the usual Lipschitz and linear growth conditions. We denote by $\mathbb{P}^{X^{t,x}}$ the measure induced by this solution. Given the functions $k, f : [0,T] \times \mathbb{R}^d \longrightarrow \mathbb{R}$ and $g : \mathbb{R}^d \longrightarrow \mathbb{R}$, we consider the optimal stopping problem

$$V(t,x) := \sup_{\tau \in \mathcal{T}_{[t,T]}^t} \mathbb{E}\left[\int_t^\tau \beta(t,s) f(s, X_s^{t,x})\mathrm{d}s + \beta(t,\tau) g(X_\tau^{t,x})\right], \tag{5.15}$$

whenever this expected value is well defined, where

$$\beta(t,s) := \mathrm{e}^{-\int_t^s k(r, X_r^{t,x})\mathrm{d}r},\ 0 \le t \le s \le T.$$

By the results of the previous chapter, the corresponding dynamic programming equation is

$$\min\{-\partial_t v - \mathcal{L}v - f, v - g\} = 0 \text{ on } [0,T) \times \mathbb{R}^d,\ v(T,.) = g, \tag{5.16}$$

where $\mathcal{L}$ is the second-order differential operator

$$\mathcal{L}v := b \cdot Dv + \frac{1}{2}\mathrm{Tr}\left[\sigma\sigma^{\mathrm{T}} D^2 v\right] - kv.$$

Similar to Sect. 5.1, a function v will be called a *supersolution* (resp. *subsolution*) of (5.16) if

$$\min\{-\partial_t v - \mathcal{L}v - f, v - g\} \ge \text{ (resp. } \le) \ 0 \text{ and } v(T,.) \ge \text{ (resp. } \le) \ g.$$

Before stating the main result of this section, we observe that for many interesting examples, it is known that the value function V does not satisfy the $C^{1,2}$ regularity which we have been using so far for the application of Itô's formula. Therefore, in order to state a result which can be applied to a wider class of problems, we shall enlarge in the following remark the set of function for which Itô's formula still holds true.

Remark 5.6. Let v be a function in the Sobolev space $W^{1,2}(\mathbf{S})$. By definition, for such a function v, there is a sequence of functions $(v^n)_{n\geq 1} \subset C^{1,2}(\mathbf{S})$ such that $v^n \longrightarrow v$ uniformly on compact subsets of $\mathbf{S}$, and

$$\|\partial_t v^n - \partial_t v^m\|_{\mathbb{L}^2(\mathbf{S})} + \|Dv^n - Dv^m\|_{\mathbb{L}^2(\mathbf{S})} + \|D^2 v^n - D^2 v^m\|_{\mathbb{L}^2(\mathbf{S})} \longrightarrow 0.$$

Then, Itô's formula holds true for v^n for all $n \geq 1$, and is inherited by v by sending $n \to \infty$.

Theorem 5.7. *Let $T < \infty$ and $v \in W^{1,2}([0,T), \mathbb{R}^d)$. Assume further that v and f have quadratic growth. Then:*

(i) *If v is a supersolution of* (5.16) $\mathrm{d}t_x \mathrm{d}\mathbb{P}^{X^{t,x}}$*- a.e. for all (t,x), then $v \geq V$.*
(ii) *If v is a solution of* (5.16) $\mathrm{d}t_x \mathrm{d}\mathbb{P}^{X^{t,x}}$*- a.e. for all (t,x), then $v = V$, and*

$$\tau_t^* := \inf\{s > t : v(s, X_s) = g(X_s)\}$$

is an optimal stopping time.

Proof. Let $(t,x) \in [0,T) \times \mathbb{R}^d$ be fixed and denote $\beta_s := \beta(t,s)$.

(i) For an arbitrary stopping time $\tau \in \mathcal{T}^t_{[t,T)}$, we denote

$$\tau_n := \tau \wedge \inf\left\{s > t : |X_s^{t,x} - x| > n\right\}.$$

By our regularity conditions on v, notice that Itô's formula can be applied to it piecewise. Then:

$$\begin{aligned} v(t,x) &= \beta_{\tau_n} v(\tau_n, X_{\tau_n}^{t,x}) - \int_t^{\tau_n} \beta_s (\partial_t + \mathcal{L}) v(s, X_s^{t,x}) \mathrm{d}s \\ &\quad - \int_t^{\tau_n} \beta_s (\sigma^{\mathrm{T}} Dv)(s, X_s^{t,x}) \mathrm{d}W_s \\ &\geq \beta_{\tau_n} v(\tau_n, X_{\tau_n}^{t,x}) + \int_t^{\tau_n} \beta_s f(s, X_s^{t,x}) \mathrm{d}s \\ &\quad - \int_t^{\tau_n} \beta_s (\sigma^{\mathrm{T}} Dv)(s, X_s^{t,x}) \mathrm{d}W_s \end{aligned}$$

by the supersolution property of v. Since $(s, X_s^{t,x})$ is bounded on the stochastic interval $[t, \tau_n]$, this provides

$$v(t,x) \geq \mathbb{E}\left[\beta_{\tau_n} v(\tau_n, X_{\tau_n}^{t,x}) + \int_t^{\tau_n} \beta_s f(s, X_s^{t,x}) \mathrm{d}s\right].$$

Notice that $\tau_n \longrightarrow \tau$ a.s. Then, since f and v have quadratic growth, we may pass to the limit $n \to \infty$ invoking the dominated convergence theorem, and we get:

$$v(t,x) \geq \mathbb{E}\left[\beta_T v(T, X_T^{t,x}) + \int_t^T \beta_s f(s, X_s^{t,x}) \mathrm{d}s\right].$$

Since $v(T,.) \geq g$ by the supersolution property, this concludes the proof of (i).

(ii) Let τ_t^* be the stopping time introduced in the theorem. Then, since $v(T,.) = g$, it follows that $\tau_t^* \in \mathcal{T}_{[t,T]}^t$. Set

$$\tau_t^n := \tau_t^* \wedge \inf\left\{s > t : |X_s^{t,x} - x| > n\right\}.$$

Observe that $v > g$ on $[t, \tau_t^n) \subset [t, \tau_t^*)$, and therefore $-\partial_t v - \mathcal{L}v - f = 0$ on $[t, \tau_t^n)$. Then, proceeding as in the previous step, it follows from Itô's formula that:

$$v(t,x) = \mathbb{E}\left[\beta_{\tau_t^n} v(\tau_t^n, X_{\tau_t^n}^{t,x}) + \int_t^{\tau_t^n} \beta_s f(s, X_s^{t,x}) \mathrm{d}s\right].$$

Since $\tau_t^n \longrightarrow \tau_t^*$ a.s. and f, v have quadratic growth, we may pass to the limit $n \to \infty$ invoking the dominated convergence theorem. This leads to

$$v(t,x) = \mathbb{E}\left[\beta_T v(T, X_T^{t,x}) + \int_t^T \beta_s f(s, X_s^{t,x}) \mathrm{d}s\right],$$

and the required result follows from the fact that $v(T,.) = g$. □

5.4 Examples of Optimal Stopping Problems with Explicit Solutions

5.4.1 *Perpetual American Options*

The pricing problem of perpetual American put options reduces to the infinite horizon optimal stopping problem:

$$P(t,s) := \sup_{\tau \in \mathcal{T}_{[t,\infty)}^t} \mathbb{E}\left[e^{-r(\tau - t)}(K - S_\tau^{t,s})^+\right],$$

where $K > 0$ is a given exercise price, $S^{t,s}$ is defined by the Black-Scholes constant coefficients model:

$$S_u^{t,s} := s \exp\left(r - \frac{\sigma^2}{2}\right)(u-t) + \sigma(W_u - W_t), \quad u \geq t,$$

and $r \geq 0, \sigma > 0$ are two given constants. By the time homogeneity of the problem, we see that

$$P(t,s) = P(s) := \sup_{\tau \in \mathcal{T}_{[0,\infty)}} \mathbb{E}\left[e^{-r\tau}(K - S_\tau^{0,s})^+\right]. \quad (5.17)$$

In view this time independence, it follows that the dynamic programming corresponding to this problem is:

$$\min\left\{v - (K-s)^+, rv - rsDv - \frac{1}{2}\sigma^2 D^2 v\right\} = 0. \quad (5.18)$$

In order to proceed to a verification argument, we now guess a solution to the previous obstacle problem. From the nature of the problem, we search for a solution of this obstacle problem defined by a parameter $s_0 \in (0, K)$ such that

$$p(s) = K - s \text{ for } s \in [0, s_0] \text{ and } rp - rsp' - \frac{1}{2}\sigma^2 s^2 p'' = 0 \text{ on } [s_0, \infty).$$

We are then reduced to solving a linear second-order ODE on $[s_0, \infty)$, thus determining v by

$$p(s) = As + Bs^{-2r/\sigma^2} \qquad \text{for } s \in [s_0, \infty),$$

up to the two constants A and B. Notice that $0 \leq p \leq K$. Then the constant $A = 0$ in our candidate solution because otherwise $v \longrightarrow \infty$ at infinity. We finally determine the constants B and s_0 by requiring our candidate solution to be continuous and differentiable at s^*. This provides two equations:

$$Bs_0^{-2r/\sigma^2} = K - s_0 \text{ and } \frac{-2r/\sigma^2}{B} s_0^{-2r/\sigma^2 - 1} = -1,$$

which provide our final candidate

$$s_0 = \frac{2rK}{2r + \sigma^2}, \quad p(s) = (K-s)\mathbf{1}_{[0,s_0]}(s) + \mathbf{1}_{[s_0,\infty)}(s) \frac{\sigma^2 s_0}{2r} \left(\frac{s}{s_0}\right)^{\frac{-2r}{\sigma^2}}. \quad (5.19)$$

Notice that our candidate p is not twice differentiable at s_0 as $p''(s_0-) = 0 \neq p''(s_0+)$. However, by Remark 5.6, Itô's formula still applies to p, and p satisfies the dynamic programming equation (5.18). We now show that

$$p = P \text{ with optimal stopping time } \tau^* := \inf\{t > 0 : p(S_t^{0,s}) = (K - S_t^{0,s})^+\}. \tag{5.20}$$

Indeed, for an arbitrary stopping time $\tau \in \mathcal{T}_{[0,\infty)}$, it follows from Itô's formula that:

$$\begin{aligned} p(s) &= e^{-r\tau} p(S_\tau^{0,s}) - \int_0^\tau e^{-rt} \left(-rp + rsp' + \frac{1}{2}\sigma^2 s^2 p''\right)(S_t)dt \\ &\quad - \int_0^\tau p'(S_t)\sigma S_t dW_t \\ &\geq e^{-r\tau}(K - S_\tau^{t,s})^+ - \int_0^\tau p'(S_t)\sigma S_t dW_t \end{aligned}$$

by the fact that p is a supersolution of the dynamic programming equation. Since p' is bounded, there is no need any localization to get rid of the stochastic integral, and we directly obtain by taking expected values that $p(s) \geq \mathbb{E}[e^{-r\tau}(K - S_\tau^{t,s})^+]$. By the arbitrariness of $\tau \in \mathcal{T}_{[0,\infty)}$, this shows that $p \geq P$.

We next repeat the same argument with the stopping time τ^*, and we see that $p(s) = \mathbb{E}[e^{-r\tau^*}(K - S_{\tau^*}^{0,s})^+]$, completing the proof of (5.20).

5.4.2 *Finite Horizon American Options*

Finite horizon optimal stopping problems rarely have an explicit solution. So, the following example can be seen as a sanity check. In the context of the financial market of the previous subsection, we assume the instantaneous interest rate $r = 0$, and we consider an American option with payoff function g and maturity $T > 0$. Then the price of the corresponding American option is given by the optimal stopping problem:

$$P(t, s) := \sup_{\tau \in \mathcal{T}_{[t,T]}^t} \mathbb{E}\big[g(S_\tau^{t,s})\big]. \tag{5.21}$$

The corresponding dynamic programming equation is:

$$\min\left\{v - g, -\partial_t v - \frac{1}{2}D^2 v\right\} = 0 \text{ on } [0, T) \times \mathbb{R}_+ \text{ and } v(T, .) = g. \tag{5.22}$$

Assuming further that $g \in W^{1,2}$ and concave, we see that g is a solution of the dynamic programming equation. Then, provided that g satisfies suitable growth condition, we see by a verification argument that $P = p$.

Notice that the previous result can be obtained directly by the Jensen inequality together with the fact that S is a martingale.

Chapter 6
Introduction to Viscosity Solutions

Throughout this chapter, we provide the main tools from the theory of viscosity solutions for the purpose of our applications to stochastic control problems. For a deeper presentation, we refer to the excellent overview paper by Crandall et al. [14].

6.1 Intuition Behind Viscosity Solutions

We consider a non-linear second order degenerate partial differential equation

$$\text{(E)}\quad F\left(x, u(x), Du(x), D^2u(x)\right) = 0 \quad \text{for} \quad x \in \mathcal{O},$$

where $\mathcal{O}$ is an open subset of $\mathbb{R}^d$ and F is a continuous map from $\mathcal{O} \times \mathbb{R} \times \mathbb{R}^d \times \mathcal{S}_d \longrightarrow \mathbb{R}$. We shall denote by $d(\mathcal{O})$ the closure of $\mathcal{O}$ in $\mathbb{R}^d$. A crucial condition on F is the so-called *ellipticity* condition:

Standing Assumption. For all $(x, r, p) \in \mathcal{O} \times \mathbb{R} \times \mathbb{R}^d$ and $A, B \in \mathcal{S}_d$,

$$F(x, r, p, A) \le F(x, r, p, B) \text{ whenever } A \ge B.$$

The full importance of this condition will be made clear in Proposition 6.2 below.

The first step towards the definition of a notion of weak solution to (E) is the introduction of sub-and supersolutions.

Definition 6.1. A function $u : \mathcal{O} \longrightarrow \mathbb{R}$ is a classical supersolution (resp. subsolution) of (E) if $u \in C^2(\mathcal{O})$ and

$$F\left(x, u(x), Du(x), D^2u(x)\right) \ge \text{ (resp. } \le) \ 0 \quad \text{for} \quad x \in \mathcal{O}.$$

The theory of viscosity solutions is motivated by the following result, whose simple proof is left to the reader.

N. Touzi, *Optimal Stochastic Control, Stochastic Target Problems, and Backward SDE*, Fields Institute Monographs 29, DOI 10.1007/978-1-4614-4286-8_6,

Proposition 6.2. *Let u be a $C^2(\mathcal{O})$ function. Then the following claims are equivalent:*

(i) *u is a classical supersolution (resp. subsolution) of (E).*
(ii) *For all pairs $(x_0, \varphi) \in \mathcal{O} \times C^2(\mathcal{O})$ such that x_0 is a minimizer (resp. maximizer) of the difference $u - \varphi$) on $\mathcal{O}$, we have*

$$F\left(x_0, u(x_0), D\varphi(x_0), D^2\varphi(x_0)\right) \geq (\textit{resp.} \leq)\ 0.$$

6.2 Definition of Viscosity Solutions

For the convenience of the reader, we recall the definition of the semicontinuous envelopes. For a locally bounded function $u : \mathcal{O} \longrightarrow \mathbb{R}$, we denote by u_* and u^* the lower- and upper-semicontinuous envelopes of u. We recall that u_* is the largest lower-semicontinuous minorant of u, u^* is the smallest upper-semicontinuous majorant of u, and

$$u_*(x) = \liminf_{x' \to x} u(x'), \quad u^*(x) = \limsup_{x' \to x} u(x') .$$

We are now ready for the definition of viscosity solutions. Observe that Claim (ii) in the above proposition does not involve the regularity of u. It therefore suggests the following weak notion of solution to (E).

Definition 6.3. Let $u : \mathcal{O} \longrightarrow \mathbb{R}$ be a locally bounded function:

(i) We say that u is a (discontinuous) viscosity supersolution of (E) if

$$F\left(x_0, u_*(x_0), D\varphi(x_0), D^2\varphi(x_0)\right) \geq 0$$

for all pairs $(x_0, \varphi) \in \mathcal{O} \times C^2(\mathcal{O})$ such that x_0 is a minimizer of the difference $(u_* - \varphi)$ on $\mathcal{O}$.
(ii) We say that u is a (discontinuous) viscosity subsolution of (E) if

$$F\left(x_0, u^*(x_0), D\varphi(x_0), D^2\varphi(x_0)\right) \leq 0$$

for all pairs $(x_0, \varphi) \in \mathcal{O} \times C^2(\mathcal{O})$ such that x_0 is a maximizer of the difference $(u^* - \varphi)$ on $\mathcal{O}$.
(iii) We say that u is a (discontinuous) viscosity solution of (E) if it is both a viscosity supersolution and subsolution of (E).

Notation. We will say that $F\left(x, u_*(x), Du_*(x), D^2u_*(x)\right) \geq 0$ in the viscosity sense whenever u_* is a viscosity supersolution of (E). A similar notation will be used for subsolution.

Remark 6.4. An immediate consequence of Proposition 6.2 is that any classical solution of (E) is also a viscosity solution of (E).

Remark 6.5. Clearly, the above definition is not changed if the minimum or maximum are local and/or strict. Also, by a density argument, the test function can be chosen in $C^\infty(\mathcal{O})$.

Remark 6.6. Consider the equation (E^+): $|u'(x)| - 1 = 0$ on $\mathbb{R}$. Then:

- The function $f(x) := |x|$ is not a viscosity supersolution of (E^+). Indeed, the test function $\varphi \equiv 0$ satisfies $(f - \varphi)(0) = 0 \leq (f - \varphi)(x)$ for all $x \in \mathbb{R}$. But $|\varphi'(0)| = 0 \not\geq 1$.
- The function $g(x) := -|x|$ is a viscosity solution of (E^+). To see this, we concentrate on the origin which is the only critical point. The supersolution property is obviously satisfied as there is no smooth function which satisfies the minimum condition. As for the subsolution property, we observe that whenever $\varphi \in C^1(\mathbb{R})$ satisfies $(g-\varphi)(0) = \max(g-\varphi)$, then $|\varphi'(0)| \leq 1$, which is exactly the viscosity subsolution property of g.
- Similarly, the function f is a viscosity solution of the equation (E^-): $-|u'(x)| + 1 = 0$ on $\mathbb{R}$.

In Sect. 7.1, we will show that the value function V is a viscosity solution of the DPE (3.21) under the conditions of Theorem 3.6 (except the smoothness assumption on V). We also want to emphasize that proving that the value function is a viscosity solution is almost as easy as proving that it is a classical solution when V is known to be smooth.

6.3 First Properties

We now turn to two important properties of viscosity solutions: the change of variable formula and the stability result.

Proposition 6.7. *Let u be a locally bounded (discontinuous) viscosity supersolution of (E). If f is a $C^1(\mathbb{R})$ function with $Df \neq 0$ on $\mathbb{R}$, then the function $v := f^{-1} \circ u$ is a (discontinuous)*

- *Viscosity supersolution, when $Df > 0$*
- *Viscosity subsolution, when $Df < 0$*

of the equation

$$K(x, v(x), Dv(x), D^2v(x)) = 0 \quad \text{for } x \in \mathcal{O},$$

where

$$K(x, r, p, A) := F\left(x, f(r), Df(r)p, D^2 f(r)pp' + Df(r)A\right).$$

We leave the easy proof of this proposition to the reader. The next result shows how limit operations with viscosity solutions can be performed very easily.

Theorem 6.8. *Let u_ε be a lower-semicontinuous viscosity supersolution of the equation*

$$F_\varepsilon\left(x, u_\varepsilon(x), Du_\varepsilon(x), D^2u_\varepsilon(x)\right) = 0 \text{ for } x \in \mathcal{O},$$

where $(F_\varepsilon)_{\varepsilon>0}$ is a sequence of continuous functions satisfying the ellipticity condition. Suppose that $(\varepsilon, x) \longmapsto u_\varepsilon(x)$ and $(\varepsilon, z) \longmapsto F_\varepsilon(z)$ are locally bounded, and define

$$\underline{u}(x) := \liminf_{(\varepsilon,x')\to(0,x)} u_\varepsilon(x') \text{ and } \overline{F}(z) := \limsup_{(\varepsilon,z')\to(0,z)} F_\varepsilon(z').$$

Then, $\underline{u}$ is a lower-semicontinuous viscosity supersolution of the equation

$$\overline{F}\left(x, \underline{u}(x), D\underline{u}(x), D^2\underline{u}(x)\right) = 0 \quad \text{for } x \in \mathcal{O}.$$

A similar statement holds for subsolutions.

Proof. The fact that $\underline{u}$ is a lower-semicontinuous function is left as an exercise for the reader. Let $\varphi \in C^2(\mathcal{O})$ and $\bar{x}$ be a strict minimizer of the difference $\underline{u} - \varphi$. By definition of $\underline{u}$, there is a sequence $(\varepsilon_n, x_n) \in (0, 1] \times \mathcal{O}$ such that

$$(\varepsilon_n, x_n) \longrightarrow (0, \bar{x}) \text{ and } u_{\varepsilon_n}(x_n) \longrightarrow \underline{u}(\bar{x}).$$

Consider some $r > 0$ together with the closed ball $\bar{B}$ with radius r, centered at $\bar{x}$. Of course, we may choose $|x_n - \bar{x}| < r$ for all $n \geq 0$. Let $\bar{x}_n$ be a minimizer of $u_{\varepsilon_n} - \varphi$ on $\bar{B}$. We claim that

$$\bar{x}_n \longrightarrow \bar{x} \text{ and } u_{\varepsilon_n}(\bar{x}_n) \longrightarrow \underline{u}(\bar{x}) \text{ as } n \to \infty. \tag{6.1}$$

Before verifying this, let us complete the proof. We first deduce that $\bar{x}_n$ is an interior point of $\bar{B}$ for large n, so that $\bar{x}_n$ is a local minimizer of the difference $u_{\varepsilon_n} - \varphi$. Then

$$F_{\varepsilon_n}\left(\bar{x}_n, u_{\varepsilon_n}(\bar{x}_n), D\varphi(\bar{x}_n), D^2\varphi(\bar{x}_n)\right) \geq 0,$$

and the required result follows by taking limits and using the definition of $\overline{F}$.

It remains to prove claim (6.1). Recall that $(x_n)_n$ is valued in the compact set $\bar{B}$. Then, there is a subsequence, still named $(x_n)_n$, which converges to some $\tilde{x} \in \bar{B}$. We now prove that $\tilde{x} = \bar{x}$ and obtain the second claim in (6.1) as a by-product. Using the fact that $\bar{x}_n$ is a minimizer of $u_{\varepsilon_n} - \varphi$ on $\bar{B}$, together with the definition of $\underline{u}$, we see that

$$
\begin{aligned}
0 = (\underline{u} - \varphi)(\bar{x}) &= \lim_{n\to\infty} \, (u_{\varepsilon_n} - \varphi)\,(x_n) \\
&\geq \limsup_{n\to\infty} \, (u_{\varepsilon_n} - \varphi)\,(\bar{x}_n) \\
&\geq \liminf_{n\to\infty} \, (u_{\varepsilon_n} - \varphi)\,(\bar{x}_n) \\
&\geq (\underline{u} - \varphi)(\tilde{x}) \, .
\end{aligned}
$$

We now obtain (6.1) from the fact that $\bar{x}$ is a strict minimizer of the difference $(\underline{u} - \varphi)$. □

Observe that the passage to the limit in partial differential equations written in the classical or the generalized sense usually requires much more technicalities, as one has to ensure convergence of all the partial derivatives involved in the equation. The above stability result provides a general method to pass to the limit when the equation is written in the viscosity sense, and its proof turns out to be remarkably simple.

A possible application of the stability result is to establish the convergence of numerical schemes. In view of the simplicity of the above statement, the notion of viscosity solutions provides a nice framework for such questions. This issue will be studied later in Chap. 12.

The main difficulty in the theory of viscosity solutions is the interpretation of the equation in the viscosity sense. First, by weakening the notion of solution to the second-order nonlinear PDE (E), we are enlarging the set of solutions, and one has to guarantee that uniqueness still holds (in some convenient class of functions). This issue will be discussed in the subsequent Sect. 6.4. We conclude this section by the following result whose proof is trivial in the classical case, but needs some technicalities when stated in the viscosity sense.

Proposition 6.9. *Let $A \subset \mathbb{R}^{d_1}$ and $B \subset \mathbb{R}^{d_2}$ be two open subsets, and let $u : A \times B \longrightarrow \mathbb{R}$ be a lower semicontinuous viscosity supersolution of the equation:*

$$
F\left(x, y, u(x,y), D_y u(x,y), D_y^2 u(x,y)\right) \geq 0 \text{ on } A \times B,
$$

where F is a continuous elliptic operator. Then, for all fixed $x_0 \in A$, the function $v(y) := u(x_0, y)$ is a viscosity supersolution of the equation:

$$
F\left(x_0, y, v(y), Dv(y), D^2 v(y)\right) \geq 0 \text{ on } B.
$$

A similar statement holds for the subsolution property.

Proof. Fix $x_0 \in A$, set $v(y) := u(x_0, y)$, and let $y_0 \in B$ and $f \in C^2(B)$ be such that

$$
(v - f)(y_0) < (v - f)(y) \quad \text{for all } y \in J \setminus \{y_0\}, \tag{6.2}
$$

where J is an arbitrary compact subset of B containing y_0 in its interior. For each integer n, define

$$\varphi_n(x,y) \;:=\; f(y) - n|x-x_0|^2 \quad \text{for } (x,y) \in A \times B,$$

and let (x_n, y_n) be defined by

$$(u-\varphi_n)(x_n,y_n) \;=\; \min_{I\times J}(u-\varphi_n),$$

where I is a compact subset of A containing x_0 in its interior. We claim that

$$(x_n,y_n) \longrightarrow (x_0,y_0) \text{ and } u(x_n,y_n) \longrightarrow u(x_0,y_0) \;\text{ as } n \longrightarrow \infty. \tag{6.3}$$

Before proving this, let us complete the proof. Since (x_0, y_0) is an interior point of $A \times B$, it follows from the viscosity property of u that

$$\begin{aligned} 0 &\le F\left(x_n, y_n, u(x_n,y_n), D_y\varphi_n(x_n,y_n), D_y^2\varphi_n(x_n,y_n)\right) \\ &= F\left(x_n, y_n, u(x_n,y_n), Df(y_n), D^2 f(y_n)\right), \end{aligned}$$

and the required result follows by sending n to infinity.

We now turn to the proof of (6.3). Since the sequence $(x_n, y_n)_n$ is valued in the compact subset $A \times B$, we have $(x_n, y_n) \longrightarrow (\bar{x}, \bar{y}) \in A \times B$, after passing to a subsequence. Observe that

$$\begin{aligned} u(x_n,y_n) - f(y_n) &\le u(x_n,y_n) - f(y_n) + n|x_n - x_0|^2 \\ &= (u-\varphi_n)(x_n,y_n) \\ &\le (u-\varphi_n)(x_0,y_0) \;=\; u(x_0,y_0) - f(y_0)\,. \end{aligned}$$

Taking the limits, it follows from the lower semicontinuity of u that

$$\begin{aligned} u(\bar{x},\bar{y}) - f(\bar{y}) &\le \liminf_{n\to\infty} u(x_n,y_n) - f(y_n) + n|x_n - x_0|^2 \\ &\le \limsup_{n\to\infty} u(x_n,y_n) - f(y_n) + n|x_n - x_0|^2 \\ &\le u(x_0,y_0) - f(y_0). \end{aligned} \tag{6.4}$$

Since u is lower semicontinuous, this implies that $u(\bar{x},\bar{y}) - f(\bar{y}) + \liminf_{n\to\infty} n|x_n - x_0|^2 \le u(x_0,y_0) - f(y_0)$. Then, we must have $\bar{x} = x_0$, and

$$(v-f)(\bar{y}) \;=\; u(x_0,\bar{y}) - f(\bar{y}) \le (v-f)(y_0),$$

which implies that $\bar{y} = y_0$ in view of (6.2), and $n|x_n - x_0|^2 \longrightarrow 0$. We also deduce from inequalities (6.4) that $u(x_n, y_n) \longrightarrow u(x_0, y_0)$, concluding the proof of (6.3). □

6.4 Comparison Result and Uniqueness

In this section, we show that the notion of viscosity solutions is consistent with the maximum principle for a wide class of equations. Once we will have such a result, the reader must be convinced that the notion of viscosity solutions is a good weakening of the notion of classical solution.

We recall that the maximum principle is a stronger statement than uniqueness, i.e., any equation satisfying a comparison result has no more than one solution.

In the viscosity solutions literature, the maximum principle is rather called *comparison principle*.

6.4.1 Comparison of Classical Solutions in a Bounded Domain

Let us first review the maximum principle in the simplest classical sense.

Proposition 6.10. *Assume that $\mathcal{O}$ is an open bounded subset of $\mathbb{R}^d$, and the nonlinearity $F(x,r,p,A)$ is elliptic and strictly increasing in r. Let $u, v \in C^2(\mathrm{cl}(\mathcal{O}))$ be classical subsolution and supersolution of* (E)*, respectively, with $u \le v$ on $\partial\mathcal{O}$. Then $u \le v$ on* $\mathrm{cl}(\mathcal{O})$.

Proof. Our objective is to prove that

$$M := \sup_{\mathrm{cl}(\mathcal{O})}(u-v) \le 0.$$

Assume to the contrary that $M > 0$. Then since $\mathrm{cl}(\mathcal{O})$ is a compact subset of $\mathbb{R}^d$, and $u - v \le 0$ on $\partial\mathcal{O}$, we have

$$M = (u-v)(x_0) \text{ for some } x_0 \in \mathcal{O} \text{ with } D(u-v)(x_0) = 0,\ D^2(u-v)(x_0) \le 0. \tag{6.5}$$

Then, it follows from the viscosity properties of u and v that:

$$\begin{aligned} F\big(x_0, u(x_0), Du(x_0), D^2u(x_0)\big) \le 0 &\le F\big(x_0, v(x_0), Dv(x_0), D^2v(x_0)\big) \\ &\le F\big(x_0, u(x_0) - M, Du(x_0), D^2u(x_0)\big), \end{aligned}$$

where the last inequality follows crucially from the ellipticity of F. This provides the desired contradiction, under our condition that F is strictly increasing in r. □

The objective of this section is to mimic the previous proof in the sense of viscosity solutions.

6.4.2 Semijets Definition of Viscosity Solutions

We first need to develop a convenient alternative definition of viscosity solutions. For $x_0 \in \mathcal{O}$, $r \in \mathbb{R}$, $p \in \mathbb{R}^d$, and $A \in \mathcal{S}_d$, we introduce the quadratic function:

$$q(y, r, p, A) := r + p \cdot y + \frac{1}{2} Ay \cdot y, \quad y \in \mathbb{R}^d.$$

For $v \in \mathrm{LSC}(\mathcal{O})$, let $(x_0, \varphi) \in \mathcal{O} \times C^2(\mathcal{O})$ be such that x_0 is a local minimizer of the difference $(v - \varphi)$ in $\mathcal{O}$. Then, defining $p := D\varphi(x_0)$ and $A := D^2\varphi(x_0)$, it follows from a second-order Taylor expansion that:

$$v(x) \geq q\big(x - x_0, v(x_0), p, A\big) + \circ\big(|x - x_0|^2\big).$$

Motivated by this observation, we introduce the *subjet* $J^-_{\mathcal{O}} v(x_0)$ by

$$J^-_{\mathcal{O}} v(x_0) := \Big\{(p, A) \in \mathbb{R}^d \times \mathcal{S}_d : v(x) \geq q\big(x - x_0, v(x_0), p, A\big) + \circ\big(|x - x_0|^2\big)\Big\}. \tag{6.6}$$

Similarly, we define the *superjet* $J^+_{\mathcal{O}} u(x_0)$ of a function $u \in \mathrm{USC}(\mathcal{O})$ at the point $x_0 \in \mathcal{O}$ by

$$J^+_{\mathcal{O}} u(x_0) := \Big\{(p, A) \in \mathbb{R}^d \times \mathcal{S}_d : u(x) \leq q\big(x - x_0, u(x_0), p, A\big) + \circ\big(|x - x_0|^2\big)\Big\}. \tag{6.7}$$

Then, it can prove that a function $v \in \mathrm{LSC}(\mathcal{O})$ is a viscosity supersolution of the equation (E) if and only if

$$F(x, v(x), p, A) \geq 0 \quad \text{for all } (p, A) \in J^-_{\mathcal{O}} v(x).$$

The nontrivial implication of the previous statement requires to construct, for every $(p, A) \in J^-_{\mathcal{O}} v(x_0)$, a smooth test function φ such that the difference $(v - \varphi)$ has a local minimum at x_0. We refer to Fleming and Soner [20], Lemma V.4.1 p. 211.

A symmetric statement holds for viscosity subsolutions. By continuity considerations, we can even enlarge the semijets $J^{\pm}_{\mathcal{O}} w(x_0)$ to the following closure:

$$\bar{J}^{\pm}_{\mathcal{O}} w(x) := \Big\{(p, A) \in \mathbb{R}^d \times \mathcal{S}_d : (x_n, w(x_n), p_n, A_n) \longrightarrow (x, w(x), p, A)$$
$$\text{for some sequence } (x_n, p_n, A_n)_n \subset \mathrm{Graph}(J^{\pm}_{\mathcal{O}} w)\Big\},$$

where $(x_n, p_n, A_n) \in \mathrm{Graph}(J^{\pm}_{\mathcal{O}} w)$ means that $(p_n, A_n) \in J^{\pm}_{\mathcal{O}} w(x_n)$. The following result is obvious, and provides an equivalent definition of viscosity solutions.

Proposition 6.11. *Consider an elliptic nonlinearity F, and let $u \in \mathrm{USC}(\mathcal{O})$, $v \in \mathrm{LSC}(\mathcal{O})$:*

(i) *Assume that F is lower semicontinuous. Then, u is a viscosity subsolution of (E) if and only if*

$$F(x, u(x), p, A) \leq 0 \quad \textit{for all } x \in \mathcal{O} \textit{ and } (p, A) \in \bar{J}^+_{\mathcal{O}} u(x).$$

(ii) *Assume that F is upper semicontinuous. Then, v is a viscosity supersolution of (E) if and only if*

$$F(x, v(x), p, A) \geq 0 \quad \textit{for all } x \in \mathcal{O} \textit{ and } (p, A) \in \bar{J}^-_{\mathcal{O}} v(x).$$

6.4.3 The Crandall–Ishii's Lemma

The major difficulty in mimicking the proof of Proposition 6.10 is to derive an analogous statement to (6.5) without involving the smoothness of u and v, as these functions are only known to be upper- and lowern semicontinuous in the context of viscosity solutions.

This is provided by the following result due to M. Crandall and I. Ishii. For a symmetric matrix, we denote by $|A| := \sup\{(A\xi) \cdot \xi : |\xi| \leq 1\}$.

Lemma 6.12. *Let $\mathcal{O}$ be an open subset of $\mathbb{R}^d$. Given $u \in USC(\mathcal{O})$ and $v \in LSC(\mathcal{O})$, we assume for some $(x_0, y_0) \in \mathcal{O}^2$, $\varphi \in C^2\left(cl(\mathcal{O})^2\right)$ that:*

$$(u - v - \varphi)(x_0, y_0) = \max_{\mathcal{O}^2}(u - v - \varphi). \tag{6.8}$$

Then, for each $\varepsilon > 0$, there exist $A, B \in \mathcal{S}_d$ such that

$$(D_x\varphi(x_0, y_0), A) \in \bar{J}^+_{\mathcal{O}} u(x_0), \quad (-D_y\varphi(x_0, y_0), B) \in \bar{J}^-_{\mathcal{O}} v(y_0),$$

and the following inequality holds in the sense of symmetric matrices in $\mathcal{S}_{2d}$:

$$-\left(\varepsilon^{-1} + \left|D^2\varphi(x_0, y_0)\right|\right) I_{2d} \leq \begin{pmatrix} A & 0 \\ 0 & -B \end{pmatrix} \leq D^2\varphi(x_0, y_0) + \varepsilon D^2\varphi(x_0, y_0)^2.$$

Proof. See Sect. 6.7. □

We will be applying Lemma 6.12 in the particular case

$$\varphi(x, y) := \frac{\alpha}{2}|x - y|^2 \text{ for } x, y \in \mathcal{O}. \tag{6.9}$$

Intuitively, sending α to ∞, we expect that the maximization of $(u(x) - v(y) - \varphi(x, y)$ on $\mathcal{O}^2$ reduces to the maximization of $(u - v)$ on $\mathcal{O}$ as in (6.5). Then, taking $\varepsilon^{-1} = \alpha$, we directly compute that the conclusions of Lemma 6.12 reduce to

$$(\alpha(x_0 - y_0), A) \in \bar{J}^+_{\mathcal{O}} u(x_0), \quad (\alpha(x_0 - y_0), B) \in \bar{J}^-_{\mathcal{O}} v(y_0), \tag{6.10}$$

and

$$-3\alpha \begin{pmatrix} I_d & 0 \\ 0 & I_d \end{pmatrix} \leq \begin{pmatrix} A & 0 \\ 0 & -B \end{pmatrix} \leq 3\alpha \begin{pmatrix} I_d & -I_d \\ -I_d & I_d \end{pmatrix}. \tag{6.11}$$

Remark 6.13. If u and v were C^2 functions in Lemma 6.12, the first-and-second order condition for the maximization problem (6.8) with the test function (6.9) is $Du(x_0) = \alpha(x_0 - y_0)$, $Dv(x_0) = \alpha(x_0 - y_0)$, and

$$\begin{pmatrix} D^2u(x_0) & 0 \\ 0 & -D^2v(y_0) \end{pmatrix} \leq \alpha \begin{pmatrix} I_d & -I_d \\ -I_d & I_d \end{pmatrix}.$$

Hence, the right-hand side inequality in (6.11) is worsening the previous second-order condition by replacing the coefficient α by 3α.

Remark 6.14. The right-hand side inequality of (6.11) implies that

$$A \leq B. \tag{6.12}$$

To see this, take an arbitrary $\xi \in \mathbb{R}^d$, and denote by ξ^{T} its transpose. From right-hand side inequality of (6.11), it follows that

$$0 \geq (\xi^{\mathrm{T}}, \xi^{\mathrm{T}}) \begin{pmatrix} A & 0 \\ 0 & -B \end{pmatrix} \begin{pmatrix} \xi \\ \xi \end{pmatrix} = (A\xi) \cdot \xi - (B\xi) \cdot \xi.$$

6.4.4 Comparison of Viscosity Solutions in a Bounded Domain

We now prove a comparison result for viscosity sub- and supersolutions by using Lemma 6.12 to mimic the proof of Proposition 6.10. The statement will be proved under the following conditions on the nonlinearity F which will be used at Step 3 of the subsequent proof.

Assumption 6.15. (i) *There exists $\gamma > 0$ such that*

$$F(x, r, p, A) - F(x, r', p, A) \geq \gamma(r - r') \textit{ for all } r \geq r', \ (x, p, A) \in \mathcal{O} \times \mathbb{R}^d \times \mathcal{S}_d.$$

(ii) *There is a function $\varpi : \mathbb{R}_+ \longrightarrow \mathbb{R}_+$ with $\varpi(0+) = 0$, such that*

$$F(y,r,\alpha(x-y),B) - F(x,r,\alpha(x-y),A) \le \varpi\left(\alpha|x-y|^2 + |x-y|\right)$$
$$\textit{for all } x,y \in \mathcal{O},\ r \in \mathbb{R}, \textit{and } A,B \textit{ satisfying } (6.11).$$

Remark 6.16. Assumption 6.15(ii) implies that the nonlinearity F is elliptic. To see this, notice that for $A \le B$, $\xi,\eta \in \mathbb{R}^d$, and $\varepsilon > 0$, we have

$$\begin{aligned} A\xi\cdot\xi - (B+\varepsilon I_d)\eta\cdot\eta &\le B\xi\cdot\xi - (B+\varepsilon I_d)\eta\cdot\eta \\ &= 2\eta\cdot B(\xi-\eta) + B(\xi-\eta)\cdot(\xi-\eta) - \varepsilon|\eta|^2 \\ &\le \varepsilon^{-1}|B(\xi-\eta)|^2 + B(\xi-\eta)\cdot(\xi-\eta) \\ &\le |B|\left(1+\varepsilon^{-1}|B|\right)|\xi-\eta|^2. \end{aligned}$$

For $3\alpha \ge (1+\varepsilon^{-1}|B|)|B|$, the latter inequality implies the right-hand side of (6.11) holds true with $(A, B+\varepsilon I_d)$. For ε sufficiently small, the left-hand side of (6.11) is also true with $(A, B+\varepsilon I_d)$ if in addition $\alpha > |A| \vee |B|$. Then

$$F(x-\alpha^{-1}p, r, p, B+\varepsilon I) - F(x,r,p,A) \le \varpi\left(\alpha^{-1}(|p|^2+|p|)\right),$$

which provides the ellipticity of F by sending $\alpha \to \infty$ and $\varepsilon \to 0$.

Theorem 6.17. *Let $\mathcal{O}$ be an open bounded subset of $\mathbb{R}^d$ and let F be an elliptic operator satisfying Assumption 6.15. Let $u \in USC(\mathcal{O})$ and $v \in LSC(\mathcal{O})$ be viscosity subsolution and supersolution of the equation (E), respectively. Then*

$$u \le v \text{ on } \partial\mathcal{O} \Longrightarrow u \le v \text{ on } \bar{\mathcal{O}} := cl(\mathcal{O}).$$

Proof. As in the proof of Proposition 6.10, we assume to the contrary that

$$\delta := (u-v)(z) > 0 \quad \text{for some } z \in \mathcal{O}. \tag{6.13}$$

Step 1. For every $\alpha > 0$, it follows from the upper semicontinuity of the difference $(u-v)$ and the compactness of $\bar{\mathcal{O}}$ that

$$\begin{aligned} M_\alpha &:= \sup_{\mathcal{O}\times\mathcal{O}} \left\{u(x) - v(y) - \frac{\alpha}{2}|x-y|^2\right\} \\ &= u(x_\alpha) - v(y_\alpha) - \frac{\alpha}{2}|x_\alpha - y_\alpha|^2 \end{aligned} \tag{6.14}$$

for some $(x_\alpha, y_\alpha) \in \bar{\mathcal{O}}\times\bar{\mathcal{O}}$. Since $\bar{\mathcal{O}}$ is compact, there is a subsequence $(x_n, y_n) := (x_{\alpha_n}, y_{\alpha_n})$, $n \ge 1$, which converges to some $(\hat{x},\hat{y}) \in \bar{\mathcal{O}}\times\bar{\mathcal{O}}$. We shall prove in Step 4 below that

$$\hat{x} = \hat{y},\ \alpha_n|x_n - y_n|^2 \longrightarrow 0, \text{ and } M_{\alpha_n} \longrightarrow (u-v)(\hat{x}) = \sup_{\mathcal{O}}(u-v). \tag{6.15}$$

Then, since $u \le v$ on $\partial\mathcal{O}$ and

$$\delta \le M_{\alpha_n} = u(x_n) - v(y_n) - \frac{\alpha_n}{2}|x_n - y_n|^2 \qquad (6.16)$$

by (6.13), it follows from the first claim in (6.15) that $(x_n, y_n) \in \mathcal{O} \times \mathcal{O}$.

Step 2. Since the maximizer (x_n, y_n) of M_{α_n} defined in (6.14) is an interior point to $\mathcal{O} \times \mathcal{O}$, it follows from Lemma 6.12 that there exist two symmetric matrices $A_n, B_n \in \mathcal{S}_n$ satisfying (6.11) such that $(x_n, \alpha_n(x_n - y_n), A_n) \in \bar{J}^+_{\mathcal{O}}u(x_n)$ and $(y_n, \alpha_n(x_n - y_n), B_n) \in \bar{J}^-_{\mathcal{O}}v(y_n)$. Then, since u and v are viscosity subsolution and supersolution, respectively, it follows from the alternative definition of viscosity solutions in Proposition 6.11 that

$$F(x_n, u(x_n), \alpha_n(x_n - y_n), A_n) \le 0 \le F(y_n, v(y_n), \alpha_n(x_n - y_n), B_n). \qquad (6.17)$$

Step 3. We first use the strict monotonicity Assumption 6.15(i) to obtain

$$\begin{aligned}\gamma\delta \ \le\ \gamma\big(u(x_n) - v(y_n)\big) \le F(x_n, u(x_n), \alpha_n(x_n - y_n), A_n)\\ -F(x_n, v(y_n), \alpha_n(x_n - y_n), A_n).\end{aligned}$$

By (6.17), this provides

$$\gamma\delta \le F(y_n, v(y_n), \alpha_n(x_n - y_n), B_n) - F(x_n, v(y_n), \alpha_n(x_n - y_n), A_n).$$

Finally, in view of Assumption 6.15(ii) this implies that

$$\gamma\delta \le \varpi\left(\alpha_n|x_n - y_n|^2 + |x_n - y_n|\right).$$

Sending n to infinity, this leads to the desired contradiction of (6.13) and (6.15).

Step 4. It remains to prove the claims (6.15). By the upper semicontinuity of the difference $(u - v)$ and the compactness of $\bar{\mathcal{O}}$, there exists a maximizer x^* of the difference $(u - v)$. Then

$$(u - v)(x^*) \le M_{\alpha_n} = u(x_n) - v(y_n) - \frac{\alpha_n}{2}|x_n - y_n|^2.$$

Sending $n \to \infty$, this provides

$$\begin{aligned}\bar{\ell} := \frac{1}{2}\limsup_{n\to\infty} \alpha_n|x_n - y_n|^2 &\le \limsup_{n\to\infty} u(x_{\alpha_n}) - v(y_{\alpha_n}) - (u - v)(x^*)\\ &\le u(\hat{x}) - v(\hat{y}) - (u - v)(x^*);\end{aligned}$$

in particular, $\bar{\ell} < \infty$ and $\hat{x} = \hat{y}$. Using the definition of x^* as a maximizer of $(u-v)$, we see that:

$$0 \leq \bar{\ell} \leq (u-v)(\hat{x}) - (u-v)(x^*) \leq 0.$$

Then $\hat{x}$ is a maximizer of the difference $(u-v)$ and $M_{\alpha_n} \longrightarrow \sup_{\mathcal{O}}(u-v)$. □

We list below two interesting examples of operators F which satisfy the conditions of the above theorem:

Example 6.18. Assumption 6.15 is satisfied by the nonlinearity

$$F(x,r,p,A) = \gamma r + H(p)$$

for any continuous function $H : \mathbb{R}^d \longrightarrow \mathbb{R}$, and $\gamma > 0$.

In this example, the condition $\gamma > 0$ is not needed when H is a convex and $H(D\varphi(x)) \leq \alpha < 0$ for some $\varphi \in C^1(\mathcal{O})$. This result can be found in [2].

Example 6.19. Assumption 6.15 is satisfied by

$$F(x,r,p,A) = -\mathrm{Tr}\left(\sigma\sigma'(x)A\right) + \gamma r,$$

where $\sigma : \mathbb{R}^d \longrightarrow \mathcal{S}_d$ is a Lipschitz function, and $\gamma > 0$. Condition (i) of Assumption 6.15 is obvious. To see that condition (ii) is satisfied, we consider $(A, B, \alpha) \in \mathcal{S}_d \times \mathcal{S}_d \times \mathbb{R}_+^*$ satisfying (6.11). We claim that

$$\mathrm{Tr}\left[MM^{\mathrm{T}}A - NN^{\mathrm{T}}B\right] \leq 3\alpha|M-N|^2 = 3\alpha\sum_{i,j=1}^{d}(M-N)_{ij}^2.$$

To see this, observe that the matrix

$$C := \begin{pmatrix} NN^{\mathrm{T}} & NM^{\mathrm{T}} \\ MN^{\mathrm{T}} & MM^{\mathrm{T}} \end{pmatrix}$$

is a non-negative matrix in $\mathcal{S}_d$. From the right hand side inequality of (6.11), this implies that

$$\begin{aligned}
\mathrm{Tr}\left[MM^{\mathrm{T}}A - NN^{\mathrm{T}}B\right] &= \mathrm{Tr}\left[C\begin{pmatrix} A & 0 \\ 0 & -B \end{pmatrix}\right] \\
&\leq 3\alpha\mathrm{Tr}\left[C\begin{pmatrix} I_d & -I_d \\ -I_d & I_d \end{pmatrix}\right] \\
&= 3\alpha\mathrm{Tr}\left[(M-N)(M-N)^{\mathrm{T}}\right] = 3\alpha|M-N|^2.
\end{aligned}$$

6.5 Comparison in Unbounded Domains

When the domain $\mathcal{O}$ is unbounded, a growth condition on the functions u and v is needed. Then, by using the growth at infinity, we can build on the proof of Theorem 6.17 to obtain a comparison principle. The following result shows how to handle this question in the case of a sub-quadratic growth. We emphasize that the present argument can be adapted to alternative growth conditions.

The following condition differs from Assumption 6.15 only in its part (ii) where the constant 3 in (6.11) is replaced by 4 in (6.18). Thus, the following Assumption 6.20(ii) is slightly stronger than Assumption 6.15(ii).

Assumption 6.20. (i) *There exists $\gamma > 0$ such that*

$$F(x,r,p,A)-F(x,r',p,A) \geq \gamma(r-r') \textit{ for all } r \geq r',\ (x,p,A) \in \mathcal{O}\times\mathbb{R}^d\times\mathcal{S}_d.$$

(ii) *There is a function $\varpi : \mathbb{R}_+ \longrightarrow \mathbb{R}_+$ with $\varpi(0+) = 0$, such that*

$$F(y,r,\alpha(x-y),B) - F(x,r,\alpha(x-y),A) \leq \varpi\left(\alpha|x-y|^2 + |x-y|\right)$$
$$\textit{for all } x, y \in \mathcal{O},\ r \in \mathbb{R} \textit{ and } A, B \textit{ satisfying}$$
$$-4\alpha\begin{pmatrix} I_d & 0 \\ 0 & I_d \end{pmatrix} \leq \begin{pmatrix} A & 0 \\ 0 & -B \end{pmatrix} \leq 4\alpha\begin{pmatrix} I_d & -I_d \\ -I_d & I_d \end{pmatrix}. \tag{6.18}$$

Theorem 6.21. *Let F be a uniformly continuous elliptic operator satisfying Assumption 6.20. Let $u \in USC(\mathcal{O})$ and $v \in LSC(\mathcal{O})$ be viscosity subsolution and supersolution of the equation (E), respectively, with $|u(x)| + |v(x)| = \circ(|x|^2)$ as $|x| \to \infty$. Then*

$$u \leq v \textit{ on } \partial\mathcal{O} \Longrightarrow u \leq v \textit{ on } cl(\mathcal{O}).$$

Proof. We assume to the contrary that

$$\delta := (u-v)(z) > 0 \quad \text{for some } z \in \mathbb{R}^d, \tag{6.19}$$

and we work towards a contradiction. Let

$$M_\alpha := \sup_{x,y\in\mathbb{R}^d} u(x) - v(y) - \phi(x,y),$$

where

$$\phi(x,y) := \frac{1}{2}\left(\alpha|x-y|^2 + \varepsilon|x|^2 + \varepsilon|y|^2\right).$$

1. Since $u(x) = \circ(|x|^2)$ and $v(y) = \circ(|y|^2)$ at infinity, there is a maximizer (x_α, y_α) for the previous problem:

$$M_\alpha = u(x_\alpha) - v(y_\alpha) - \phi(x_\alpha, y_\alpha).$$

Moreover, there is a sequence $\alpha_n \to \infty$ such that

$$(x_n, y_n) := (x_{\alpha_n}, y_{\alpha_n}) \longrightarrow (\hat{x}, \hat{y}),$$

and, similar to Step 4 of the proof of Theorem 6.17, we can prove that $\hat{x} = \hat{y}$,

$$\alpha_n |x_n - y_n|^2 \longrightarrow 0, \text{ and } M_{\alpha_n} \longrightarrow M_\infty := \sup_{x \in \mathbb{R}^d} (u - v)(x) - \varepsilon |x|^2. \quad (6.20)$$

Notice that

$$\begin{aligned}
\limsup_{n \to \infty} M_{\alpha_n} &= \limsup_{n \to \infty} \{u(x_n) - v(y_n) - \phi(x_n, y_n)\} \\
&\le \limsup_{n \to \infty} \{u(x_n) - v(y_n))\} \\
&\le \limsup_{n \to \infty} u(x_n) - \liminf_{n \to \infty} v(y_n) \\
&\le (u - v)(\hat{x}).
\end{aligned}$$

Since $u \le v$ on $\partial \mathcal{O}$, and

$$M_{\alpha_n} \ge \delta - \varepsilon |z|^2 > 0,$$

by (6.19), we deduce that $\hat{x} \notin \partial \mathcal{O}$ and therefore (x_n, y_n) is a local maximizer of $u - v - \phi$.

2. By the Crandall–Ishii Lemma 6.12, there exist $A_n, B_n \in \mathcal{S}_n$, such that

$$\begin{aligned}
&(D_x \phi(x_n, y_n), A_n) \in \bar{\mathcal{J}}_{\mathcal{O}}^{2,+} u(t_n, x_n), \\
&\big(-D_y \phi(x_n, y_n), B_n\big) \in \bar{\mathcal{J}}_{\mathcal{O}}^{2,-} v(t_n, y_n),
\end{aligned} \quad (6.21)$$

and

$$\begin{aligned}
-(\alpha + |D^2 \phi(x_0, y_0)|) I_{2d} &\le \begin{pmatrix} A_n & 0 \\ 0 & -B_n \end{pmatrix} \\
&\le D^2 \phi(x_n, y_n) + \frac{1}{\alpha} D^2 \phi(x_n, y_n)^2. \quad (6.22)
\end{aligned}$$

In the present situation, we immediately calculate that

$$D_x \phi(x_n, y_n) = \alpha(x_n - y_n) + \varepsilon x_n, , \quad - D_y \phi(x_n, y_n) = \alpha(x_n - y_n) - \varepsilon y_n$$

and

$$D^2\phi(x_n, y_n) = \alpha \begin{pmatrix} I_d & -I_d \\ -I_d & I_d \end{pmatrix} + \varepsilon\, I_{2d},$$

which reduces the right hand side of (6.22) to

$$\begin{pmatrix} A_n & 0 \\ 0 & -B_n \end{pmatrix} \leq (3\alpha + 2\varepsilon) \begin{pmatrix} I_d & -I_d \\ -I_d & I_d \end{pmatrix} + \left(\varepsilon + \frac{\varepsilon^2}{\alpha}\right) I_{2d}, \qquad (6.23)$$

while the left land side of (6.22) implies

$$-3\alpha I_{2d} \leq \begin{pmatrix} A_n & 0 \\ 0 & -B_n \end{pmatrix}. \qquad (6.24)$$

3. By (6.21) and the viscosity properties of u and v, we have

$$F(x_n, u(x_n), \alpha_n(x_n - y_n) + \varepsilon x_n, A_n) \leq 0,$$
$$F(y_n, v(y_n), \alpha_n(x_n - y_n) - \varepsilon y_n, B_n) \geq 0.$$

Using Assumption 6.20(i) together with the uniform continuity of H, this implies that

$$\gamma\big(u(x_n) - v(x_n)\big) \leq F\big(y_n, u(x_n), \alpha_n(x_n - y_n), \tilde{B}_n\big)$$
$$-F\big(x_n, u(x_n), \alpha_n(x_n - y_n), \tilde{A}_n\big) + c(\varepsilon),$$

where $c(.)$ is a modulus of continuity of F, and $\tilde{A}_n := A_n - 2\varepsilon I_n$, $\tilde{B}_n := B_n + 2\varepsilon I_n$. By (6.23) and (6.24), we have

$$-4\alpha I_{2d} \leq \begin{pmatrix} \tilde{A}_n & 0 \\ 0 & -\tilde{B}_n \end{pmatrix} \leq 4\alpha \begin{pmatrix} I_d & -I_d \\ -I_d & I_d \end{pmatrix},$$

for small ε. Then, it follows from Assumption 6.20(ii) that

$$\gamma\big(u(x_n) - v(x_n)\big) \leq \varpi\big(\alpha_n|x_n - y_n|^2 + |x_n - y_n|\big) + c(\varepsilon).$$

By sending n to infinity, it follows from (6.20) that:

$$c(\varepsilon) \geq \gamma\big(M_\infty + |\hat{x}|^2\big) \geq \gamma M_\infty \geq \gamma\big(u(z) - v(z) - \varepsilon|z|^2\big),$$

and we get a contradiction of (6.19) by sending ε to zero. □

6.6 Useful Applications

We conclude this section by two consequences of the above comparison results, which are trivial properties in the context of classical solutions.

Lemma 6.22. *Let $\mathcal{O}$ be an open interval of $\mathbb{R}$, and $U : \mathcal{O} \longrightarrow \mathbb{R}$ be a lower-semicontinuous viscosity supersolution of the equation $DU \geq 0$ on $\mathcal{O}$. Then U is nondecreasing on $\mathcal{O}$.*

Proof. For each $\varepsilon > 0$, define $W(x) := U(x) + \varepsilon x$, $x \in \mathcal{O}$. Then W satisfies in the viscosity sense $DW \geq \varepsilon$ in $\mathcal{O}$, i.e., for all $(x_0, \varphi) \in \mathcal{O} \times C^1(\mathcal{O})$ such that

$$(W - \varphi)(x_0) = \min_{x \in \mathcal{O}} (W - \varphi)(x), \tag{6.25}$$

we have $D\varphi(x_0) \geq \varepsilon$. This proves that φ is strictly increasing in a neighborhood $\mathcal{V}$ of x_0. Let $(x_1, x_2) \subset \mathcal{V}$ be an open interval containing x_0. We intend to prove that

$$W(x_1) < W(x_2), \tag{6.26}$$

which provides the required result from the arbitrariness of $x_0 \in \mathcal{O}$.

To prove (6.26), suppose to the contrary that $W(x_1) \geq W(x_2)$, and the consider the function $v(x) = W(x_2)$ which solves the equation

$$Dv = 0 \text{ on the open interval } (x_1, x_2).$$

Together with the boundary conditions $v(x_1) = v(x_2) = W(x_2)$. Observe that W is a lower-semicontinuous viscosity supersolution of the above equation. From the comparison result of Example 6.18, this implies that

$$\sup_{[x_1, x_2]} (v - W) = \max\{(v - W)(x_1), (v - W)(x_2)\} \leq 0.$$

Hence $W(x) \geq v(x) = W(x_2)$ for all $x \in [x_1, x_2]$. Applying this inequality at $x_0 \in (x_1, x_2)$ and recalling that the test function φ is strictly increasing on $[x_1, x_2]$, we get

$$(W - \varphi)(x_0) > (W - \varphi)(x_2),$$

contradicting (6.25). □

Lemma 6.23. *Let $\mathcal{O}$ be an open interval of $\mathbb{R}$, and $U : \mathcal{O} \longrightarrow \mathbb{R}$ be a lower-semicontinuous viscosity supersolution of the equation $-D^2U \geq 0$ on $\mathcal{O}$. Then U is concave on $\mathcal{O}$.*

Proof. Let $a < b$ be two arbitrary elements in $\mathcal{O}$, and consider some $\varepsilon > 0$ together with the function

$$v(s) := \frac{U(a)\left(e^{\sqrt{\varepsilon}(b-s)} - e^{-\sqrt{\varepsilon}(b-s)}\right) + U(b)\left(e^{\sqrt{\varepsilon}(s-a)} - e^{-\sqrt{\varepsilon}(s-a)}\right)}{e^{\sqrt{\varepsilon}(b-a)} - e^{-\sqrt{\varepsilon}(b-a)}} \text{ for } a \le s \le b.$$

Clearly, v solves the equation

$$\varepsilon v - D^2 v = 0 \ \text{ on } \ (a,b), \ v = U \ \text{ on } \ \{a,b\}.$$

Since U is lower semicontinuous, it is bounded from below on the interval $[a,b]$. Therefore, by possibly adding a constant to U, we can assume that $U \ge 0$, so that U is a lower-semicontinuous viscosity supersolution of the above equation. It then follows from the comparison Theorem 7.6 that:

$$\sup_{[a,b]} (v - U) = \max\{(v-U)(a), (v-U)(b)\} \ = \ 0.$$

Hence,

$$U(s) \ge v(s) \ = \ \frac{U(a)\left(e^{\sqrt{\varepsilon}(b-s)} - e^{-\sqrt{\varepsilon}(b-s)}\right) + U(b)\left(e^{\sqrt{\varepsilon}(s-a)} - e^{-\sqrt{\varepsilon}(s-a)}\right)}{e^{\sqrt{\varepsilon}(b-a)} - e^{-\sqrt{\varepsilon}(b-a)}}$$

and by sending ε to zero, we see that

$$U(s) \ge (U(b) - U(a))\frac{s-a}{b-a} + U(a)$$

for all $s \in [a,b]$. Let λ be an arbitrary element of the interval [0,1], and set $s := \lambda a + (1-\lambda)b$. The last inequality takes the form

$$U\big(\lambda a + (1-\lambda)b\big) \ge \lambda U(a) + (1-\lambda)U(b),$$

proving the concavity of U. □

6.7 Proof of the Crandall–Ishii's Lemma

We start with two lemmas. We say that a function f is $\lambda-$semiconvex if $x \longmapsto f(x) + (\lambda/2)|x|^2$ is convex.

Lemma 6.24. *Let $f : \mathbb{R}^N \longrightarrow \mathbb{R}$ be a $\lambda-$semiconvex function, for some $\lambda \in \mathbb{R}$, and assume that $f(x) - \frac{1}{2}Bx \cdot x \le f(0)$ for all $x \in \mathbb{R}^N$. Then there exists $X \in \mathcal{S}_N$ such that*

$$(0, X) \in \overline{J}^{2,+} f(0) \cap \overline{J}^{2,-} f(0) \text{ and } -\lambda I_N \le X \le B.$$

Our second lemma requires to introduce the following notion. For a function $v : \mathbb{R}^N \longrightarrow \mathbb{R}$ and $\lambda > 0$, the corresponding λ*–sup-convolution* is defined by:

$$\hat{v}^\lambda(x) := \sup_{y\in\mathbb{R}^N} \left\{ v(y) - \frac{\lambda}{2}|x-y|^2 \right\}.$$

Observe that

$$\hat{v}^\lambda(x) + \frac{\lambda}{2}|x|^2 = \sup_{y\in\mathbb{R}^N} \left\{ v(y) - \frac{\lambda}{2}|y|^2 + \lambda x \cdot y \right\}$$

is convex, as the supremum of linear functions. Then

$$\hat{v}^\lambda \text{ is } \lambda - \text{semiconvex.} \tag{6.27}$$

In [14], the following property is referred to as the *magical property of the sup-convolution.*

Lemma 6.25. *Let $\lambda > 0$, v be a bounded lower-semicontinuous function and $\hat{v}^\lambda$ the corresponding λ–sup-convolution.*

(i) *If $(p, X) \in J^{2,+}\hat{v}(x)$ for some $x \in \mathbb{R}^N$, then*

$$(p, X) \in J^{2,+}v\Big(x + \frac{p}{\lambda}\Big) \text{ and } \hat{v}^\lambda(x) = v(x + p/\lambda) - \frac{1}{2\lambda}|p|^2.$$

(ii) *For all $x \in \mathbb{R}^N$, we have $(0, X) \in \bar{J}^{2,+}\hat{v}(x)$implies that $(0, X) \in \bar{J}^{2,+}v(x)$.*

Before proving the above lemmas, we show how they imply the Crandall–Ishii's lemma that we reformulate in a more symmetric way.

Lemma 6.12. *Let $\mathcal{O}$ be an open locally compact subset of $\mathbb{R}^d$ and $u_1, u_2 \in USC(\mathcal{O})$. We denote $w(x_1, x_2) := u_1(x_1) + u_2(x_2)$ and we assume for some $\varphi \in C^2\big(cl(\mathcal{O})^2\big)$, and $x^0 = (x_1^0, x_2^0) \in \mathcal{O} \times \mathcal{O}$ that*

$$(w - \varphi)(x^0) = \max_{\mathcal{O}^2}(w - \varphi).$$

Then, for each $\varepsilon > 0$, there exist $X_1, X_2 \in \mathcal{S}_d$ such that

$$(D_{x_i}\varphi(x^0), X_i) \in \bar{J}^{2,+}_{\mathcal{O}} u_i(x_i^0), \quad i = 1, 2,$$

and

$$-\left(\varepsilon^{-1} + \left|D^2\varphi(x^0)\right|\right) I_{2d} \leq \begin{pmatrix} X_1 & 0 \\ 0 & X_2 \end{pmatrix} \leq D^2\varphi(x^0) + \varepsilon D^2\varphi(x^0)^2.$$

Proof.

Step 1. We first observe that we may reduce the problem to the case

$$\mathcal{O} = \mathbb{R}^d,\ x^0 = 0,\ u_1(0) = u_2(0) = 0, \text{ and } \varphi(x) = \frac{1}{2}Ax \cdot x \text{ for some } A \in \mathcal{S}_d.$$

The reduction to $x^0 = 0$ follows from an immediate change of coordinates. Choose any compact subset of $K \subset \mathcal{O}$ containing the origin and set $\bar{u}_i = u_i$ on K and $-\infty$ otherwise, $i = 1, 2$. Then, the problem can be stated equivalently in terms of the functions $\bar{u}_i$ which are now defined on $\mathbb{R}^d$ and take values on the extended real line. Also by defining

$$\bar{\bar{u}}_i(x_i) := \bar{u}_i(x_i) - u_i(0) - D_{x_i}\varphi(0) \text{ and } \bar{\varphi}(x) := \varphi(x) - \varphi(0) - D\varphi(0) \cdot x,$$

we may reformulate the problem equivalently with $\bar{\bar{u}}_i(x_i) = 0$ and $\bar{\varphi}(x) = \frac{1}{2}D^2\varphi(0)x \cdot x + \circ(|x|^2)$. Finally, defining $\bar{\bar{\varphi}}(x) := Ax \cdot x$ with $A := D^2\varphi(0) + \eta I_{2d}$ for some $\eta > 0$, it follows that

$$\begin{aligned}\bar{\bar{u}}_1(x_1) + \bar{\bar{u}}_2(x_2) - \bar{\bar{\varphi}}(x_1, x_2) &< \bar{\bar{u}}_1(x_1) + \bar{\bar{u}}_2(x_2) - \bar{\varphi}(x_1, x_2) \\ &\le \bar{\bar{u}}_1(0) + \bar{\bar{u}}_2(0) - \bar{\varphi}(0) = 0.\end{aligned}$$

Step 2. From the reduction of the previous step, we have

$$\begin{aligned}2w(x) &\le Ax \cdot x \\ &= A(x-y)\cdot(x-y)Ay \cdot y - 2Ay \cdot (y - x) \\ &\le A(x-y)\cdot(x-y)Ay \cdot y + \varepsilon A^2 y \cdot y + \frac{1}{\varepsilon}|x-y|^2 \\ &= A(x-y)\cdot(x-y) + \frac{1}{\varepsilon}|x-y|^2 + (A + \varepsilon A^2)y \cdot y \\ &\le (\varepsilon^{-1} + |A|)|x-y|^2 + (A + \varepsilon A^2)y \cdot y.\end{aligned}$$

Set $\lambda := \varepsilon^{-1} + |A|$ and $B := A + \varepsilon A^2$. The latter inequality implies the following property of the sup-convolution:

$$\hat{w}^\lambda(y) - \frac{1}{2}By \cdot y \le \hat{w}(0) \ = \ 0.$$

Step 3. Recall from (6.27) that $\hat{w}^\lambda$ is λ–semiconvex. Then, it follows from Lemma 6.24 that there exist $X \in \mathcal{S}_{2d}$ such that $(0, X) \in \overline{J}^{2,+}\hat{w}^\lambda(0) \cap \overline{J}^{2,-}\hat{w}^\lambda(0)$ and $-\lambda I_{2d} \le X \le B$. Moreover, it is immediately checked that $\hat{w}^\lambda(x_1, x_2) = \hat{u}_1^\lambda(x_1) + \hat{u}_2^\lambda(x_2)$, implying that X is block diagonal with blocs $X_1, X_2 \in \mathcal{S}_d$. Hence,

$$-(\varepsilon^{-1} + |A|)I_{2d} \leq \begin{pmatrix} X_1 & 0 \\ 0 & X_2 \end{pmatrix} \leq A + \varepsilon A^2,$$

and $(0, X_i) \in \overline{J}^{2,+} \hat{u}_i^\lambda(0)$ for $i = 1, 2$ which, by Lemma 6.25, implies that $(0, X_i) \in \overline{J}^{2,+} u_i^\lambda(0)$. □

We continue by turning to the proofs of Lemmas 6.24 and 6.25. The main tools which will be used are the following properties of any semiconvex function $\varphi : \mathbb{R}^N \longrightarrow \mathbb{R}$ whose proofs are reported in [14]:

- *Aleksandrov's lemma.* φ is twice differentiable a.e.
- *Jensen's lemma.* if x_0 is a strict maximizer of φ, then for every $r, \delta > 0$, the set

$$\left\{\bar{x} \in B(x_0, r) : x \longmapsto \varphi(x) + p \cdot x \text{ has a local maximum at } \bar{x} \text{ for some } p \in B_\delta\right\}$$

has positive measure in $\mathbb{R}^N$.

Proof of Lemma 6.24. Notice that $\varphi(x) := f(x) - \frac{1}{2}Bx \cdot x - |x|^4$ has a strict maximum at $x = 0$. Localizing around the origin, we see that φ is a semiconvex function. Then, for every $\delta > 0$, by the above Aleksandrov's and Jensen's lemmas, there exists q_δ and x_δ such that

$$q_\delta, x_\delta \in B_\delta,\ D^2\varphi(x_\delta) \text{ exists, and } \varphi(x_\delta) + q_\delta \cdot x_\delta = \text{loc-max}\{\varphi(x) + q_\delta \cdot x\}.$$

We may then write the first- and second-order optimality conditions to see that:

$$Df(x_\delta) = -q_\delta + Bx_\delta + 4|x_\delta|^3 \text{ and } D^2 f(x_\delta) \leq B + 12|x_\delta|^2.$$

Together with the λ–semiconvexity of f, this provides:

$$Df(x_\delta) = O(\delta) \text{ and } -\lambda I \leq D^2 f(x_\delta) \leq B + O(\delta^2). \tag{6.28}$$

Clearly f inherits the twice differentiability of φ at x_δ. Then

$$\left(Df(x_\delta), D^2 f(x_\delta)\right) \in J^{2,+} f(x_\delta) \cap J^{2,-} f(x_\delta),$$

and, in view of (6.28), we may send δ to zero along some subsequence and obtain a limit point $(0, X) \in \bar{J}^{2,+} f(0) \cap \bar{J}^{2,-} f(0)$. □

Proof of Lemma 6.25. (i) Since v is bounded, there is a maximizer:

$$\hat{v}^\lambda(x) = v(y) - \frac{\lambda}{2}|x - y|^2. \tag{6.29}$$

By the definition of $\hat{v}^\lambda$ and the fact that $(p, A) \in J^{2,+}\hat{v}(x)$, we have for every $x', y' \in \mathbb{R}^N$:

$$\begin{aligned} v(y') - \frac{\lambda}{2}|x' - y'|^2 &\le \hat{v}(x') \\ &\le \hat{v}(x) + p \cdot (x' - x) + \frac{1}{2}A(x' - x) \cdot (x' - x) + \circ(x' - x) \\ &= v(y) - \frac{\lambda}{2}|x - y|^2 + p \cdot (x' - x) \\ &\quad + \frac{1}{2}A(x' - x) \cdot (x' - x) + \circ(x' - x), \end{aligned} \tag{6.30}$$

where we used (6.29) in the last equality.

By first setting $x' = y' + y - x$ in (6.30), we see that:

$$v(y') \le v(y) + p \cdot (y' - y) + \frac{1}{2}A(y' - y) \cdot (y' - y) + \circ(y' - y) \text{ for all } y' \in \mathbb{R}^N,$$

which means that $(p, A) \in J^{2,+}y(y)$.

On the other hand, setting $y' = y$ in (6.30), we deduce that:

$$\lambda(x' - x) \cdot \left(\frac{x + x'}{2} + \frac{p}{\lambda} - y\right) \ge O\left(|x - x'|^2\right),$$

which implies that $y = x + \frac{p}{\lambda}$.

(ii) Consider a sequence (x_n, p_n, A_n) with $(x_n, \hat{v}^\lambda(x_n), p_n, A_n) \longrightarrow (x, \hat{v}^\lambda(x), 0, A)$ and $(p_n, A_n) \in J^{2,+}\hat{v}^\lambda(x_n)$. In view of (i) and the definition of $\bar{J}^{2,+}v(x)$, it only remains to prove that

$$v\left(x_n + \frac{p_n}{\lambda}\right) \longrightarrow v(x). \tag{6.31}$$

To see this, we use the upper semicontinuity of v together with (i) and the definition of $\hat{v}^\lambda$:

$$\begin{aligned} v(x) &\ge \limsup_n v\left(x_n + \frac{p_n}{\lambda}\right) \\ &\ge \liminf_n v\left(x_n + \frac{p_n}{\lambda}\right) \\ &= \lim_n \hat{v}^\lambda(x_n) + \frac{1}{2\lambda}|p_n|^2 \;=\; \hat{v}^\lambda(x) \;\ge\; v(x). \end{aligned}$$

□

Chapter 7
Dynamic Programming Equation in the Viscosity Sense

7.1 DPE for Stochastic Control Problems

We now turn to the stochastic control problem introduced in Sect. 3.1. The chief goal of this section is to use the notion of viscosity solutions in order to relax the smoothness condition on the value function V in the statement of Propositions 3.4 and 3.5. Notice that the following proofs are obtained by slight modification of the corresponding proofs in the smooth case.

Remark 7.1. Recall that the general theory of viscosity applies for nonlinear partial differential equations on an open domain $\mathcal{O}$. This indeed ensures that the optimizer in the definition of viscosity solutions is an interior point. In the setting of control problems with finite horizon, the time variable moves forward so that the left boundary of the time interval is not relevant. We shall then write the DPE on the domain $\mathbf{S} = [0, T) \times \mathbb{R}^d$. Although this is not an open domain, the general theory of viscosity solutions is still valid.

We first recall the setting of Sect. 3.1. We shall concentrate on the finite horizon case $T < \infty$, while keeping in mind that the infinite horizon problems are handled by exactly the same arguments. The only reason why we exclude $T = \infty$ is because we do not want to be diverted by issues related to the definition of the set of admissible controls.

Given a subset U of $\mathbb{R}^k$, we denote by $\mathcal{U}$ the set of all progressively measurable processes $\nu = \{\nu_t, t < T\}$ valued in U and by $\mathcal{U}_0 := \mathcal{U} \cap \mathbb{H}^2$. The elements of $\mathcal{U}_0$ are called admissible control processes.

The controlled state dynamics is defined by means of the functions

$$b : (t, x, u) \in \mathbf{S} \times U \longrightarrow b(t, x, u) \in \mathbb{R}^n$$

and

$$\sigma : (t, x, u) \in \mathbf{S} \times U \longrightarrow \sigma(t, x, u) \in \mathcal{M}_{\mathbb{R}}(n, d),$$

N. Touzi, *Optimal Stochastic Control, Stochastic Target Problems, and Backward SDE*, Fields Institute Monographs 29, DOI 10.1007/978-1-4614-4286-8_7,

which are assumed to be continuous and to satisfy the conditions

$$|b(t,x,u) - b(t,y,u)| + |\sigma(t,x,u) - \sigma(t,y,u)| \le K\ |x-y|, \tag{7.1}$$

$$|b(t,x,u)| + |\sigma(t,x,u)| \le K\ (1 + |x| + |u|), \tag{7.2}$$

for some constant K independent of (t,x,y,u). For each admissible control process $\nu \in \mathcal{U}_0$, the controlled stochastic differential equation:

$$dX_t = b(t, X_t, \nu_t)dt + \sigma(t, X_t, \nu_t)dW_t, \tag{7.3}$$

has a unique solution X, for all given initial data $\xi \in \mathbb{L}^2(\mathcal{F}_0, \mathbb{P})$ with

$$\mathbb{E}\left[\sup_{0\le s\le t} |X_s^\nu|^2\right] < C(1 + \mathbb{E}[|\xi|^2])e^{Ct} \quad \text{for all } t \in [0,T], \tag{7.4}$$

for some constant C. Finally, the gain functional is defined via the functions:

$$f, k\ :\ [0,T) \times \mathbb{R}^d \times U \longrightarrow \mathbb{R} \text{ and } g\ :\ \mathbb{R}^d \longrightarrow \mathbb{R},$$

which are assumed to be continuous, $\|k^-\|_\infty < \infty$, and:

$$|f(t,x,u)| + |g(x)| \le K(1 + |u| + |x|^2),$$

for some constant K independent of (t,x,u). The cost function J on $[0,T] \times \mathbb{R}^d \times \mathcal{U}$ is:

$$J(t,x,\nu) := \mathbb{E}\left[\int_t^T \beta^\nu(t,s) f(s, X_s^{t,x,\nu}, \nu_s)ds + \beta^\nu(t,T) g\left(X_T^{t,x,\nu}\right)\right], \tag{7.5}$$

when this expression is meaningful, where

$$\beta^\nu(t,s) := \exp\left(-\int_t^s k(r, X_r^{t,x,\nu}, \nu_r)dr\right),$$

and $\{X_s^{t,x,\nu}, s \ge t\}$ is the solution of (7.3) with control process ν and initial condition $X_t^{t,x,\nu} = x$. The stochastic control problem is defined by the value function:

$$V(t,x) := \sup_{\nu\in\mathcal{U}_0} J(t,x,\nu) \text{ for } (t,x) \in \mathbf{S}. \tag{7.6}$$

We recall the expression of the Hamiltonian

$$H(.,r,p,A) := \sup_{u\in U}\left(f(.,u) - k(.,u)r + b(.,u)\cdot p + \frac{1}{2}\text{Tr}\left[\sigma\sigma^{\text{T}}(.,u)A\right]\right), \tag{7.7}$$

and the second order operator associated to X and β

$$\mathcal{L}^u v := -k(.,u)v + b(.,u)\cdot Dv + \frac{1}{2}\mathrm{Tr}\left[\sigma\sigma^{\mathrm{T}}(.,u)D^2 v\right], \tag{7.8}$$

which appears naturally in the following Itô's formula valid for any smooth test function v:

$$\mathrm{d}\beta^\nu(0,t)v(t,X_t^\nu) = \beta^\nu(0,t)\Big((\partial_t + \mathcal{L}^{\nu_t})v(t,X_t^\nu)\mathrm{d}t + Dv(t,X_t^\nu)\cdot\sigma(t,X_t^\nu,\nu_t)\mathrm{d}W_t\Big).$$

Proposition 7.2. *Assume that V is locally bounded on $[0,T)\times\mathbb{R}^d$. Then, the value function V is a viscosity supersolution of the equation*

$$-\partial_t V(t,x) - H\left(t,x,V(t,x),DV(t,x),D^2V(t,x)\right) \geq 0 \tag{7.9}$$

on $[0,T)\times\mathbb{R}^d$.

Proof. Let $(t,x)\in\mathbf{S}$ and $\varphi\in C^2(\mathbf{S})$ be such that

$$0 \;=\; (V_* - \varphi)(t,x) = \min_{\mathbf{S}}\,(V_* - \varphi). \tag{7.10}$$

Let $(t_n,x_n)_n$ be a sequence in $\mathbf{S}$ such that

$$(t_n,x_n) \;\longrightarrow\; (t,x) \text{ and } V(t_n,x_n) \;\longrightarrow\; V_*(t,x).$$

Since φ is smooth, notice that

$$\eta_n \;:=\; V(t_n,x_n) - \varphi(t_n,x_n) \longrightarrow 0.$$

Next, let $u\in U$ be fixed, and consider the constant control process $\nu = u$. We shall denote by $X^n := X^{t_n,x_n,u}$ the associated state process with initial data $X^n_{t_n} = x_n$. Finally, for all $n>0$, we define the stopping time:

$$\theta_n := \inf\{s > t_n \;:\; (s-t_n, X_s^n - x_n) \notin [0,h_n)\times\alpha B\},$$

where $\alpha>0$ is some given constant, B denotes the unit ball of $\mathbb{R}^n$, and

$$h_n := \sqrt{\eta_n}\mathbf{1}_{\{\eta_n\neq 0\}} + n^{-1}\mathbf{1}_{\{\eta_n=0\}}.$$

Notice that $\theta_n \longrightarrow t$ as $n\longrightarrow\infty$.

1. From the first inequality in the dynamic programming principle of Theorem 3.3, it follows that:

$$0 \leq \mathbb{E}\left[V(t_n,x_n) - \beta(t_n,\theta_n)V_*(\theta_n,X^n_{\theta_n}) - \int_{t_n}^{\theta_n}\beta(t_n,r)f(r,X_r^n,\nu_r)\mathrm{d}r\right].$$

Now, in contrast with the proof of Proposition 3.4, the value function is not known to be smooth, and therefore we cannot apply Itô's formula to V. The main trick of this proof is to use the inequality $V_* \geq \varphi$ on $\mathbf{S}$, implied by (7.10), so that we can apply Itô's formula to the smooth test function φ:

$$0 \leq \eta_n + \mathbb{E}\left[\varphi(t_n, x_n) - \beta(t_n, \theta_n)\varphi(\theta_n, X^n_{\theta_n}) - \int_{t_n}^{\theta_n} \beta(t_n, r) f(r, X^n_r, \nu_r)\mathrm{d}r\right]$$

$$= \eta_n - \mathbb{E}\left[\int_{t_n}^{\theta_n} \beta(t_n, r)(\partial_t\varphi + \mathcal{L}^{\cdot}\varphi - f)(r, X^n_r, u)\mathrm{d}r\right]$$

$$- \mathbb{E}\left[\int_{t_n}^{\theta_n} \beta(t_n, r) D\varphi(r, X^n_r)\sigma(r, X^n_r, u)\mathrm{d}W_r\right],$$

where $\partial_t\varphi$ denotes the partial derivative with respect to t.

2. We now continue exactly along the lines of the proof of Proposition 3.5. Observe that $\beta(t_n, r) D\varphi(r, X^n_r)\sigma(r, X^n_r, u)$ is bounded on the stochastic interval $[t_n, \theta_n]$. Therefore, the second expectation on the right hand side of the last inequality vanishes, and

$$\frac{\eta_n}{h_n} - \mathbb{E}\left[\frac{1}{h_n}\int_{t_n}^{\theta_n} \beta(t_n, r)(\partial_t\varphi + \mathcal{L}^{\cdot}\varphi - f)(r, X_r, u)\mathrm{d}r\right] \geq 0.$$

We now send n to infinity. The a.s. convergence of the random value inside the expectation is easily obtained by the mean value theorem; recall that for $n \geq N(\omega)$ sufficiently large, $\theta_n(\omega) = h_n$. Since the random variable $h_n^{-1}\int_t^{\theta_n} \beta(t_n, r)(\mathcal{L}^{\cdot}\varphi - f)(r, X^n_r, u)\mathrm{d}r$ is essentially bounded, uniformly in n, on the stochastic interval $[t_n, \theta_n]$, it follows from the dominated convergence theorem that:

$$-\partial_t\varphi(t, x) - \mathcal{L}^u\varphi(t, x) - f(t, x, u) \geq 0,$$

which is the required result, since $u \in U$ is arbitrary. □

We next wish to show that V satisfies the nonlinear partial differential equation (7.9) with equality, in the viscosity sense. This is also obtained by a slight modification of the proof of Proposition 3.5.

Proposition 7.3. *Assume that the value function V is locally bounded on $\mathbf{S}$. Let the function H be finite and upper semicontinuous on $[0, T) \times \mathbb{R}^d \times \mathbb{R}^d \times \mathcal{S}_d$, and $\|k^+\|_\infty < \infty$. Then, V is a viscosity subsolution of the equation*

$$-\partial_t V(t, x) - H\left(t, x, V(t, x), DV(t, x), D^2V(t, x)\right) \leq 0 \qquad (7.11)$$

on $[0, T) \times \mathbb{R}^n$.

Proof. Let $(t_0, x_0) \in \mathbf{S}$ and $\varphi \in C^2(\mathbf{S})$ be such that

$$0 = (V^* - \varphi)(t_0, x_0) > (V^* - \varphi)(t, x) \text{ for } (t, x) \in \mathbf{S} \setminus \{(t_0, x_0)\}. \qquad (7.12)$$

In order to prove the required result, we assume to the contrary that

$$h(t_0, x_0) := \partial_t \varphi(t_0, x_0) + H\left(t_0, x_0, \varphi(t_0, x_0), D\varphi(t_0, x_0), D^2\varphi(t_0, x_0)\right) < 0$$

and work towards a contradiction.

1. Since H is upper semicontinuous, there exists an open neighborhood $\mathcal{N}_r := (t_0 - r, t_0 + r) \times rB(t_0, x_0)$ of (t_0, x_0), for some $r > 0$, such that

$$h := \partial_t \varphi + H\left(., \varphi, D\varphi, D^2\varphi\right) < 0 \text{ on } \mathcal{N}_r. \qquad (7.13)$$

Then it follows from (7.12) that

$$-2\eta e^{r\|k^+\|_\infty} := \max_{\partial \mathcal{N}_\eta} (V^* - \varphi) < 0. \qquad (7.14)$$

Next, let $(t_n, x_n)_n$ be a sequence in $\mathcal{N}_r$ such that

$$(t_n, x_n) \longrightarrow (t_0, x_0) \text{ and } V(t_n, x_n) \longrightarrow V^*(t_0, x_0).$$

Since $(V - \varphi)(t_n, x_n) \longrightarrow 0$, we can assume that the sequence (t_n, x_n) also satisfies:

$$|(V - \varphi)(t_n, x_n)| \leq \eta \quad \text{for all } n \geq 1. \qquad (7.15)$$

For an arbitrary control process $\nu \in \mathcal{U}_{t_n}$, we define the stopping time

$$\theta_n^\nu := \inf\{t > t_n : X_t^{t_n, x_n, \nu} \notin \mathcal{N}_r\},$$

and we observe that $\left(\theta_n^\nu, X_{\theta_n^\nu}^{t_n, x_n, \nu}\right) \in \partial \mathcal{N}_r$ by the pathwise continuity of the controlled process. Then, with $\beta_s^\nu := \beta^\nu(t_n, s)$, it follows from (7.14) that

$$\beta_{\theta_n^\nu}^\nu \varphi\left(\theta_n^\nu, X_{\theta_n^\nu}^{t_n, x_n, \nu}\right) \geq 2\eta + \beta_{\theta_n^\nu}^\nu V^*\left(\theta_n^\nu, X_{\theta_n^\nu}^{t_n, x_n, \nu}\right). \qquad (7.16)$$

2. Since $\beta_{t_n}^\nu = 1$, it follows from (7.15) and Itô's formula that

$$\begin{aligned} V(t_n, x_n) &\geq -\eta + \varphi(t_n, x_n) \\ &= -\eta + \mathbb{E}\left[\beta_{\theta_n^\nu}^\nu \varphi\left(\theta_n^\nu, X_{\theta_n^\nu}^{t_n, x_n, \nu}\right) - \int_{t_n}^{\theta_n^\nu} \beta_s^\nu (\partial_t + \mathcal{L}^{\nu_s}) \varphi\left(s, X_s^{t_n, x_n, \nu}\right) ds\right] \end{aligned}$$

$$\geq -\eta + \mathbb{E}\left[\beta^{\nu}_{\theta^{\nu}_n}\varphi(\theta^{\nu}_n, X^{t_n,x_n,\nu}_{\theta^{\nu}_n}) + \int_{t_n}^{\theta^{\nu}_n} \beta^{\nu}_s\big(f(.,\nu_s) - h\big)\big(s, X^{t_n,x_n,\nu}_s\big)\mathrm{d}s\right]$$

$$\geq -\eta + \mathbb{E}\left[\beta^{\nu}_{\theta^{\nu}_n}\varphi(\theta^{\nu}_n, X^{t_n,x_n,\nu}_{\theta^{\nu}_n}) + \int_{t_n}^{\theta^{\nu}_n} \beta^{\nu}_s f\left(s, X^{t_n,x_n,\nu}_s, \nu_s\right)\mathrm{d}s\right],$$

by (7.13). Using (7.16), this provides

$$V(t_n, x_n) \geq \eta + \mathbb{E}\left[\beta^{\nu}_{\theta^{\nu}_n} V^*\left(\theta^{\nu}_n, X^{t_n,x_n,\nu}_{\theta^{\nu}_n}\right) + \int_{t_n}^{\theta^{\nu}_n} \beta^{\nu}_s f\left(s, X^{t_n,x_n,\nu}_s, \nu_s\right)\mathrm{d}s\right].$$

Since $\eta > 0$ does not depend on ν, it follows from the arbitrariness of $\nu \in \mathcal{U}_{t_n}$ that latter inequality is in contradiction with the second inequality of the dynamic programming principle of Theorem 3.3. □

As a consequence of Propositions 7.3 and 7.2, we have the main result of this section:

Theorem 7.4. *Let the conditions of Propositions 7.3 and 7.2 hold. Then, the value function V is a viscosity solution of the Hamilton-Jacobi-Bellman equation*

$$-\partial_t V - H\left(\cdot, V, DV, D^2 V\right) = 0 \text{ on } \mathbf{S}. \tag{7.17}$$

The partial differential equation (7.17) has a very simple and specific dependence in the time-derivative term. Because of this, it is usually referred to as a *parabolic* equation.

In order to a obtain a characterization of the value function by means of the dynamic programming equation, the latter viscosity property needs to be complemented by a uniqueness result. This is usually obtained as a consequence of a comparison result.

In the present situation, one may verify the conditions of Theorem 6.21. For completeness, we report a comparison result which is adapted for the class of equations corresponding to stochastic control problems.

Consider the parabolic equation

$$\partial_t u + G\left(t, x, Du(t,x), D^2u(t,x)\right) \;=\; 0 \text{ on } \mathbf{S}, \tag{7.18}$$

where G is elliptic and continuous. For $\gamma > 0$, set

$$G^{+\gamma}(t,x,p,A) := \sup\left\{G(s,y,p,A) \;:\; (s,y) \in B_{\mathbf{S}}(t,x;\gamma)\right\},$$
$$G^{-\gamma}(t,x,p,A) := \inf\left\{G(s,y,p,A) \;:\; (s,y) \in B_{\mathbf{S}}(t,x;\gamma)\right\},$$

where $B_{\mathbf{S}}(t,x;\gamma)$ is the collection of elements (s,y) in $\mathbf{S}$ such that $|t-s|^2 + |x-y|^2 \leq \gamma^2$. We report, without proof, the following result from [20] (Theorem V.8.1 and Remark V.8.1).

Assumption 7.5. *The above operators satisfy*

$$\limsup_{\varepsilon \searrow 0} \left\{ G^{+\gamma_\varepsilon}(t_\varepsilon, x_\varepsilon, p_\varepsilon, A_\varepsilon) - G^{-\gamma_\varepsilon}(s_\varepsilon, y_\varepsilon, p_\varepsilon, B_\varepsilon) \right\}$$
$$\leq \text{Const}\,(|t_0 - s_0| + |x_0 - y_0|)\,[1 + |p_0| + \alpha\,(|t_0 - s_0| + |x_0 - y_0|)] \quad (7.19)$$

for all sequences $(t_\varepsilon, x_\varepsilon), (s_\varepsilon, y_\varepsilon) \in [0,T) \times \mathbb{R}^n$, $p_\varepsilon \in \mathbb{R}^n$, *and* $\gamma_\varepsilon \geq 0$ *with:*

$$((t_\varepsilon, x_\varepsilon), (s_\varepsilon, y_\varepsilon), p_\varepsilon, \gamma_\varepsilon) \longrightarrow ((t_0, x_0), (s_0, y_0), p_0, 0) \text{ as } \varepsilon \searrow 0,$$

and symmetric matrices $(A_\varepsilon, B_\varepsilon)$ *with*

$$-KI_{2n} \leq \begin{pmatrix} A_\varepsilon & 0 \\ 0 & -B_\varepsilon \end{pmatrix} \leq 2\alpha \begin{pmatrix} I_n & -I_n \\ -I_n & I_n \end{pmatrix}$$

for some α *independent of* ε.

Theorem 7.6. *Let Assumption 7.5 hold true, and let* $u \in USC(\bar{\mathbf{S}})$, *and* $v \in LSC(\bar{\mathbf{S}})$ *be viscosity subsolution and supersolution of* (7.18), *respectively. Then*

$$\sup_{\bar{\mathbf{S}}}(u - v) = \sup_{\mathbb{R}^n}(u - v)(T, \cdot).$$

A sufficient condition for (7.19) to hold is that $f(\cdot,\cdot,u)$, $k(\cdot,\cdot,u)$, $b(\cdot,\cdot,u)$, and $\sigma(\cdot,\cdot,u) \in C^1(\bar{\mathbf{S}})$ with

$$\|b_t\|_\infty + \|b_x\|_\infty + \|\sigma_t\|_\infty + \|\sigma_x\|_\infty < \infty,$$
$$|b(t,x,u)| + |\sigma(t,x,u)| \leq \text{Const}(1 + |x| + |u|);$$

see [20], Lemma V.8.1.

7.2 DPE for Optimal Stopping Problems

We first recall the optimal stopping problem considered in Sect. 4.1. For $0 \leq t \leq T \leq \infty$, the set $\mathcal{T}_{[t,T]}$ denotes the collection of all $\mathbb{F}-$stopping times with values in $[t,T]$. The state process X is defined by the SDE:

$$dX_t = \mu(t, X_t)dt + \sigma(t, X_t)dW_t, \quad (7.20)$$

where μ and σ are defined on $\bar{\mathbf{S}} := [0,T) \times \mathbb{R}^n$, take values in $\mathbb{R}^n$ and $\mathcal{S}_n$, respectively, and satisfy the usual Lipschitz and linear growth conditions so that the above SDE has a unique strong solution satisfying the integrability of Theorem 2.2.

For a measurable function $g : \mathbb{R}^n \longrightarrow \mathbb{R}$, satisfying $\mathbb{E}\left[\sup_{0\le t<T} |g(X_t)|\right] < \infty$, the gain criterion is given by:

$$J(t,x,\tau) := \mathbb{E}\left[g\left(X_\tau^{t,x}\right)\mathbf{1}_{\tau<\infty}\right] \quad \text{for all } (t,x) \in \mathbf{S},\ \tau \in \mathcal{T}_{[t,T]}. \tag{7.21}$$

Here, $X^{t,x}$ denotes the unique strong solution of (4.1) with initial condition $X_t^{t,x} = x$. Then, the optimal stopping problem is defined by

$$V(t,x) := \sup_{\tau\in\mathcal{T}_{[t,T]}} J(t,x,\tau) \quad \text{for all } (t,x) \in \mathbf{S}. \tag{7.22}$$

The next result derives the dynamic programming equation for the previous optimal stopping problem in the sense of viscosity solution, thus relaxing the $C^{1,2}$ regularity condition in the statement of Theorem 4.5. As usual, the same methodology allows to handle seemingly more general optimal stopping problems:

$$\bar{V}(t,x) := \sup_{\tau\in\mathcal{T}_{[t,T]}} \bar{J}(t,x,\tau), \tag{7.23}$$

where

$$\bar{J}(t,x,\tau) := \mathbb{E}\left[\int_t^T \beta(t,s) f(s,X_s^{t,x})ds + \beta(t,\tau) g(X_\tau^{t,x})\mathbf{1}_{\{\tau<\infty\}}\right],$$

$$\beta(t,s) := \exp\left(-\int_t^s k(u,X_u^{t,x})du\right).$$

Theorem 7.7. *Assume that V is locally bounded, and let $g : \mathbb{R}^n \longrightarrow \mathbb{R}$ be continuous. Then V is a viscosity solution of the obstacle problem*

$$\min\{-(\partial_t + \mathcal{A})V\ ,\ V - g\} \ = \ 0 \text{ on } \mathbf{S}. \tag{7.24}$$

Proof. (i) We first show that V is a viscosity supersolution. As in the proof of Theorem 4.5, the inequality $V - g \ge 0$ is obvious, and implies that $V_* \ge g$. Let $(t_0,x_0) \in \mathbf{S}$ and $\varphi \in C^2(\mathbf{S})$ be such that

$$0 = (V_* - \varphi)(t_0,x_0) = \min_{\mathbf{S}}(V_* - \varphi).$$

To prove that $-(\partial_t + \mathcal{A})\varphi(t_0,x_0) \ge 0$, we consider a sequence $(t_n,x_n)_{n\ge1} \subset [t_0-h,t_0+h]\times B$, for some small $h>0$, such that

$$(t_n,x_n) \longrightarrow (t_0,x_0) \text{ and } V(t_n,x_n) \longrightarrow V_*(t_0,x_0).$$

Let $(h_n)_n$ be a sequence of positive scalars converging to zero, to be fixed later, and introduce the stopping times:

$$\theta_{h_n}^n := \inf\left\{t > t_n : (t, X_t^{t_n,x_n}) \notin [t_0-h_n,t_0+h_n]\times B\right\}.$$

Then $\theta_{h_n} \in \mathcal{T}^t_{[t,T]}$ for sufficiently small h, and it follows from (4.10) that:

$$V(t_n, x_n) \geq \mathbb{E}\left[V_*\left(\theta^n_h, X_{\theta^n_h}\right)\right].$$

Since $V_* \geq \varphi$, and denoting $\eta_n := (V - \varphi)(t_n, x_n)$, this provides

$$\eta_n + \varphi(t_n, x_n) \geq \mathbb{E}\left[\varphi\left(\theta^n_h, X_{\theta^n_h}\right)\right] \text{ where } \eta_n \longrightarrow 0.$$

We continue by fixing

$$h_n := \sqrt{\eta_n}\mathbf{1}_{\{\eta_n \neq 0\}} + n^{-1}\mathbf{1}_{\{\eta_n = 0\}},$$

as in the proof of Proposition 7.2. Then, the rest of the proof follows exactly the line of argument of the proof of Theorem 4.5 combined with that of Proposition 7.2.

(ii) We next prove that V is a viscosity subsolution of the (7.24). Let $(t_0, x_0) \in \mathbf{S}$ and $\varphi \in C^2(\mathbf{S})$ be such that

$$0 = (V^* - \varphi)(t_0, x_0) \;=\; \text{strict} \max_{\mathbf{S}} (V^* - \varphi),$$

assume to the contrary that

$$(V^* - g)(t_0, x_0) > 0 \text{ and } -(\partial_t + \mathcal{A})\varphi(t_0, x_0) > 0,$$

and let us work towards a contradiction of the weak dynamic programming principle.

Since g is continuous, and $V^*(t_0, x_0) = \varphi(t_0, x_0)$, we may find constants $h > 0$ and $\delta > 0$ so that

$$\varphi \geq g + \delta \text{ and } -(\partial_t + \mathcal{A})\varphi \geq 0 \text{ on } \mathcal{N}_h := [t_0, t_0 + h] \times hB, \qquad (7.25)$$

where B is the unit ball centered at x_0. Moreover, since (t_0, x_0) is a strict maximizer of the difference $V^* - \varphi$,

$$-\gamma := \max_{\partial\mathcal{N}_h}(V^* - \varphi) \;<\; 0. \qquad (7.26)$$

Let (t_n, x_n) be a sequence in $\mathbf{S}$ such that

$$(t_n, x_n) \longrightarrow (t_0, x_0) \text{ and } V(t_n, x_n) \longrightarrow V^*(t_0, x_0).$$

We next define the stopping times:

$$\theta_n := \inf\left\{t > t_n : \;\left(t, X^{t_n,x_n}_t\right) \notin \mathcal{N}_h\right\},$$

and we continue as in Step 2 of the proof of Theorem 4.5. We denote $\eta_n := V(t_n, x_n) - \varphi(t_n, x_n)$, and we compute by Itô's formula that for an arbitrary stopping rule $\tau \in \mathcal{T}^t_{[t,T]}$,

$$V(t_n, x_n) = \eta_n + \varphi(t_n, x_n)$$
$$= \eta_n + \mathbb{E}\left[\varphi(\tau \wedge \theta_n, X_{\tau\wedge\theta_n}) - \int_{t_n}^{\tau\wedge\theta_n} (\partial_t + \mathcal{A})\varphi(t, X_t)\mathrm{d}t\right],$$

where diffusion term has zero expectation because the process $(t, X_t^{t_n,x_n})$ is confined to the compact subset $\mathcal{N}_h$ on the stochastic interval $[t_n, \tau \wedge \theta_n]$. Since $-(\partial_t + \mathcal{A})\varphi \geq 0$ on $\mathcal{N}_h$ by (7.25), this provides

$$V(t_n, x_n) \geq \eta_n + \mathbb{E}\left[\varphi(\tau, X_\tau)\mathbf{1}_{\{\tau<\theta_n\}} + \varphi(\theta_n, X_{\theta_n})\mathbf{1}_{\tau\geq\{\theta_n\}}+\right]$$
$$\geq \mathbb{E}\left[(g(X_\tau) + \delta)\mathbf{1}_{\{\tau<\theta_n\}} + \left(V^*(\theta_n, X_{\theta_n}) + \gamma\right)\mathbf{1}_{\{\theta_n\geq\tau\}}\right]$$
$$\geq \gamma \wedge \delta + \mathbb{E}\left[g(X_\tau)\mathbf{1}_{\{\tau<\theta_n\}} + V^*(\theta_n, X_{\theta_n})\mathbf{1}_{\{\theta_n\geq\tau\}}\right],$$

where we used the fact that $\varphi \geq g + \delta$ on $\mathcal{N}_h$ by (7.25), and $\varphi \geq V^* + \gamma$ on $\partial\mathcal{N}_h$ by (7.26). Since $\eta_n := (V - \varphi)(t_n, x_n) \longrightarrow 0$ as $n \to \infty$, and $\tau \in \mathcal{T}^t_{[t,T]}$ is arbitrary, this provides the desired contradiction of (4.9). □

7.3 A Comparison Result for Obstacle Problems

In this section, we derive a comparison result for the obstacle problem:

$$\min\left\{F(., u, \partial_t u, Du, D^2u), u - g\right\} = 0 \text{ on } [0, T) \times \mathbb{R}^d \tag{7.27}$$

The dynamic programming equation of the optimal stopping problem (7.23) corresponds to the particular case:

$$F(., u, \partial_t u, Du, D^2u) = \partial_t u + b \cdot Du + \frac{1}{2}\mathrm{Tr}\left[\sigma\sigma^{\mathrm{T}} D^2u\right] - ku + f.$$

Theorem 7.8. *Let F be a uniformly continuous elliptic operator satisfying Assumption 6.20. Let $u \in USC(\mathcal{O})$ and $v \in LSC(\mathcal{O})$ be viscosity subsolution and supersolution of the equation (7.27), respectively, with sub-quadratic growth. Then*

$$u(T, .) \leq v(T, .) \Longrightarrow u \leq v \text{ on } [0, T] \times \mathbb{R}^d.$$

Proof. This is an easy adaptation of the proof of Theorem 6.21. We adapt the same notations so that, in the present, x stands for the pair (t, x). The only difference appears at Step 3 which starts from the fact that

$$\min\{F(x_n, u(x_n), \alpha_n(x_n - y_n) + \varepsilon x_n, A_n), u(x_n) - g(x_n)\} \le 0,$$
$$\min\{F(y_n, v(y_n), \alpha_n(x_n - y_n) - \varepsilon y_n, B_n), v(y_n) - g(y_n)\} \ge 0.$$

This leads to two cases:

- Either $u(x_n) - g(x_n) \le 0$ along some subsequence. Then the inequality $v(y_n) - g(y_n) \ge 0$ leads to a contradiction of (6.19).
- Or $F(x_n, u(x_n), \alpha_n(x_n - y_y) + \varepsilon x_n, A_n) \le 0$, which can be combined with the supersolution part $F(y_n, v(y_n), \alpha_n(x_n - y_y) - \varepsilon y_n, B_n) \ge 0$ exactly as in the proof of Theorem 6.21, and leads to a contradiction of (6.19). □

Chapter 8
Stochastic Target Problems

8.1 Stochastic Target Problems

In this section, we study a special class of stochastic target problems which avoids facing some technical difficulties, but reflects in a transparent way the main ideas and arguments to handle this new class of stochastic control problems.

All of the applications that we will be presenting fall into the framework of this section. The interested readers may consult the references at the end of this chapter for the most general classes of stochastic target problems, and their geometric formulation.

8.1.1 Formulation

Let $T > 0$ be the finite time horizon and $W = \{W_t, 0 \le t \le T\}$ be a d-dimensional Brownian motion defined on a complete probability space $(\Omega, \mathcal{F}, \mathbb{P})$. We denote by $\mathbb{F} = \{\mathcal{F}_t, 0 \le t \le T\}$ the $\mathbb{P}$-augmentation of the filtration generated by W.

We assume that the control set U is a convex compact subset of $\mathbb{R}^d$ with nonempty interior, and we denote by $\mathcal{U}$ the set of all progressively measurable processes $\nu = \{\nu_t, 0 \le t \le T\}$ with values in U.

The state process is defined as follows: given the initial data $z = (x, y) \in \mathbb{R}^d \times \mathbb{R}$, an initial time $t \in [0, T]$, and a control process $\nu \in \mathcal{U}$, let the controlled process $Z^{t,z,\nu} = (X^{t,x,\nu}, Y^{t,z,\nu})$ be the solution of the stochastic differential equation:

$$\begin{aligned} dX_r^{t,x,\nu} &= \mu\left(r, X_r^{t,x,\nu}, \nu_r\right) dr + \sigma\left(r, X_r^{t,x,\nu}, \nu_r\right) dW_r, \\ dY_r^{t,z,\nu} &= b\left(r, Z_r^{t,z,\nu}, \nu_r\right) dr + \nu_r \cdot dW_r, \quad r \in (t, T), \end{aligned}$$

with initial data

$$X_t^{t,x,\nu} = x, \text{ and } Y_t^{t,x,y,\nu} = y.$$

N. Touzi, *Optimal Stochastic Control, Stochastic Target Problems, and Backward SDE*, Fields Institute Monographs 29, DOI 10.1007/978-1-4614-4286-8_8,

Here, $\mu : \mathbf{S}\times U \longrightarrow \mathbb{R}^d$, $\sigma : \mathbf{S}\times U \longrightarrow \mathcal{S}_d$, and $b : \mathbf{S}\times\mathbb{R}\times U \longrightarrow \mathbb{R}$ are continuous functions, Lipschitz in (x, y) uniformly in (t, u). Then, all above processes are well defined for every admissible control process $\nu \in \mathcal{U}_0$ defined by

$$\mathcal{U}_0 := \left\{ \nu \in \mathcal{U} : \mathbb{E}\left[\int_0^t \left(|\mu_0(s,\nu_s)| + |b_0(s,\nu_s)| + |\sigma_0(s,\nu_s)|^2 + |\nu_s|^2\right) \mathrm{d}s\right] < \infty \right\},$$

where $\mu_0(t,u) := \mu(t,0,u)$, $b_0(t,u) := b(t,0,u)$, and $\sigma_0(t,u) := \sigma(t,0,u)$. Throughout this section, we assume that the function

$$u \longmapsto \sigma(t,x,u)p$$

has a unique fixed point for every $(t,x) \in \bar{\mathbf{S}} \times \mathbb{R}$ defined by

$$\sigma(t,x,u)p = u \Longleftrightarrow u = \psi(t,x,p). \tag{8.1}$$

For a measurable function $g : \mathbb{R}^d \longrightarrow \mathbb{R}$, we define the *stochastic target* problem by

$$V(t,x) := \inf\left\{ y \in \mathbb{R} \; : \; Y_T^{t,x,y,\nu} \geq g\left(X_T^{t,x,\nu}\right), \; \mathbb{P}-\text{a.s. for some } \nu \in \mathcal{U}_0 \right\}. \tag{8.2}$$

Remark 8.1. By introducing the subset of control processes

$$\mathcal{A}(t,x,y) := \left\{ \nu \in \mathcal{U}_0 : \; Y_T^{t,x,y,\nu} \geq g\left(X_T^{t,x,\nu}\right), \; \mathbb{P}-\text{a.s.} \right\},$$

we may re-write the value function of the stochastic target problem into

$$V(t,x) = \inf \mathcal{Y}(t,x), \text{ where } \mathcal{Y}(t,x) := \{ y \in \mathbb{R} \; : \; \mathcal{A}(t,x,y) \neq \emptyset \}.$$

The set $\mathcal{Y}(t,x)$ satisfies the following important property

$$\text{for all } \; y \in \mathbb{R}, \; y \in \mathcal{Y}(t,x) \implies [y,\infty) \subset \mathcal{Y}(t,x).$$

Indeed, since the state process $X^{t,x,\nu}$ is independent of y, the process $Y^{t,x,y,\nu}$ is a solution of a stochastic differential equation (with random coefficients), and the corresponding flow $y \longmapsto Y_t^{t,x,y,\nu}$ is increasing for every t by classical results on SDEs.

8.1.2 Geometric Dynamic Programming Principle

Similar to the standard class of stochastic control and optimal stopping problems studied in the previous chapters, the main tool for the characterization of the value function and the solution of stochastic target problems is the dynamic programming principle. Although the present problem does not fall into the class of problems studied in the previous chapters, the idea of dynamic programming is the same: allow the time origin to move, and deduce a relation between the value function at different points in time.

In these notes, we shall essentially use the *easy* direction of a more general geometric dynamic programming principle. The geometric nature of this result will be justified in Remark 8.4.

Theorem 8.2. *Let $(t,x) \in [0,T] \times \mathbb{R}^d$ and $y \in \mathbb{R}$ such that $\mathcal{A}(t,x,y) \neq \emptyset$. Then, for any control process $\nu \in \mathcal{A}(t,x,y)$ and stopping time $\tau \in \mathcal{T}_{[t,T]}$,*

$$Y_\tau^{t,x,y,\nu} \geq V\left(\tau, X_\tau^{t,x,\nu}\right), \quad P - a.s. \tag{8.3}$$

Proof. Let $z = (x,y)$ and $\nu \in \mathcal{A}(t,z)$, and denote $Z_{t,z,\nu} := (X^{t,x,\nu}, Y^{t,z,\nu})$. Then $Y_T^{t,z,\nu} \geq g\left(X_T^{t,x,\nu}\right)$ $\mathbb{P}$–a.s. Notice that

$$Z_T^{t,z,\nu} = Z_T^{\tau, Z_\tau^{t,z,\nu}, \nu}.$$

Then, by the definition of the set $\mathcal{A}$, it follows that $\nu \in \mathcal{A}\left(\tau, Z_\tau^{t,z,\nu}\right)$, and therefore $V\left(\tau, X_\tau^{t,x,\nu}\right) \leq Y_\tau^{t,z,\nu}$, $\mathbb{P}$–a.s. □

In the next subsection, we will prove that the value function V is a viscosity supersolution of the corresponding dynamic programming equation which will be obtained as the infinitesimal counterpart of (8.3). The following remark comments on the full geometric dynamic programming principle in the context of stochastic target problems. The proof of this claim is beyond the scope of these notes, and we shall only report a sketch of it.

Remark 8.3. The statement (8.3) in Theorem 8.2 can be strengthened to the following geometric dynamic programming principle:

$$V(t,x) = \inf\left\{y \in \mathbb{R} : Y_\tau^{t,x,y,\nu} \geq V\left(\tau, X_\tau^{t,x,\nu}\right), \ \mathbb{P} - \text{a.s. for some } \nu \in \mathcal{U}_0\right\}, \tag{8.4}$$

for all stopping time τ with values in $[t,T]$. Let us provide an intuitive justification of this. Denote $\hat{y} := V(t,x)$. In view of (8.3), it is easily seen that (8.4) is implied by

$$\mathbb{P}\left[Y_\tau^{t,x,\hat{y}-\eta,\nu} > V\left(\tau, X_\tau^{t,x,\nu}\right)\right] < 1 \qquad \text{for all } \nu \in \mathcal{U}_0 \text{ and } \eta > 0.$$

In words, there is no control process ν which allows to reach the value function $V(\tau, X_\tau^{t,x,\nu})$ at time τ, with full probability, starting from an initial data strictly below the value function $V(t,x)$. To see this, suppose to the contrary that there exist $\nu \in \mathcal{U}_0$, $\eta > 0$ and $\tau \in \mathcal{T}_{[t,T]}$ such that:

$$Y_\tau^{t,x,\hat{y}-\eta,\nu} > V\left(\tau, X_\tau^{t,x,\nu}\right), \quad P - \text{a.s.}$$

In view of Remark 8.1, this implies that $Y_\tau^{t,x,\hat{y}-\eta,\nu} \in \mathcal{Y}\left(\tau, X_\tau^{t,x,\nu}\right)$, and therefore there exists a control $\hat{\nu} \in \mathcal{U}_0$ such that

$$Y_T^{\tau,Z_\tau^{t,x,\hat{y}-\eta,\nu},\hat{\nu}} \geq g\left(X_T^{\tau,X_\tau^{t,x,\nu},\hat{\nu}}\right), \quad \mathbb{P}-\text{a.s.}$$

Since the process $\left(X^{\tau,X_\tau^{t,x,\nu},\hat{\nu}}, Y^{\tau,Z_\tau^{t,x,\hat{y}-\eta,\nu},\hat{\nu}}\right)$ depends on $\hat{\nu}$ only through its realizations in the stochastic interval $[\tau, T]$, we may chose $\hat{\nu}$ so as $\hat{\nu} = \nu$ on $[t, \tau]$ (this is the difficult part of this proof). Then $Z_T^{\tau,Z_\tau^{t,x,\hat{y}-\eta,\nu},\hat{\nu}} = Z_T^{t,x,\hat{y}-\eta,\hat{\nu}}$ and therefore $\hat{y} - \eta \in \mathcal{Y}(t, x)$, hence $V(t, x) \leq \hat{y} - \eta$. This is the required contradiction as $\hat{y} = V(t, x)$ and $\eta > 0$.

Remark 8.4. An extended version of the stochastic target problem that was introduced in [36] avoids the decoupling of the components $Z = (X, Y)$. In this case, there is no natural direction to isolate in the process Z which we assume defined by the general dynamics:

$$\mathrm{d}Z_r^{t,z,\nu} = \beta(r, Z_r^{t,z,\nu}, \nu_r)\mathrm{d}r + \beta(r, Z_r^{t,z,\nu}, \nu_r)\mathrm{d}W_r, \quad r \in (t, T).$$

The stochastic target problem is defined by the value function:

$$V(t) := \left\{z \in \mathbb{R}^{d+1} : Z_T^{t,z,\nu} \in \Gamma, \ \mathbb{P}-\text{a.s.}\right\},$$

for some given target subset $\Gamma \subset \mathbb{R}^{d+1}$. Notice that $V(t)$ is a subset in $\mathbb{R}^{d+1}$. It was proved in [36] that for all stopping time τ with values in $[t, T]$:

$$V(t) = \inf\left\{z \in \mathbb{R}^{d+1} : Z_\tau^{t,z,\nu} \in V(\tau), \ \mathbb{P}-\text{a.s.}\right\}.$$

This is a dynamic programming principle for the sets $\{V(t), t \in [0, T]\}$, and for this reason it was called *geometric* dynamic programming principle.

8.1.3 The Dynamic Programming Equation

In order to have a simpler statement and proof of the main result, we assume in this section that

$$U \text{ is a closed convex subset of } \mathbb{R}^d, \ \text{int}(U) \neq \emptyset \text{ and } 0 \in U. \tag{8.5}$$

The formulation of the dynamic programming equation involves the notion of support function from convex analysis.

Dual Characterization of Closed Convex Sets We first introduce the *support function* of the set U:

$$\delta_U(\xi) := \sup_{x \in U} x \cdot \xi, \quad \text{for all } \xi \in \mathbb{R}^d.$$

By definition δ_U is a convex function on $\mathbb{R}^d$. Since $0 \in U$, its effective domain is given by

$$\tilde{U} := \operatorname{dom}(\delta_U) = \{\xi \in \mathbb{R}^d \ : \ \delta_U(\xi) < \infty\},$$

and is a closed convex cone of $\mathbb{R}^d$. Since U is closed and convex by (8.5), we have the following dual characterization:

$$x \in U \text{ if and only if } \delta_U(\xi) - x \cdot \xi \geq 0 \quad \text{for all } \xi \in \tilde{U}, \tag{8.6}$$

see, e.g., Rockafellar [35]. Moreover, since $\tilde{U}$ is a cone, we may normalize the dual variables ξ on the right hand side:

$$x \in U \text{ if and only if } \delta_U(\xi) - x \cdot \xi \geq 0 \text{ for all } \xi \in \tilde{U}_1 := \{\xi \in \tilde{U} : |\xi| = 1\}. \tag{8.7}$$

This normalization will be needed in our analysis in order to obtain a dual characterization of $\operatorname{int}(U)$. Indeed, since U has nonempty interior by (8.5), we have

$$x \in \operatorname{int}(U) \quad \text{if and only if } \inf_{\xi \in \tilde{U}_1} \delta_U(\xi) - x \cdot \xi > 0. \tag{8.8}$$

Formal Derivation of the DPE We start with a formal derivation of the dynamic programming equation which provides the main intuitions.

To simplify the presentation, we suppose that the value function V is smooth and that existence holds, i.e., for all $(t, x) \in \mathbf{S}$, there is a control process $\hat{\nu} \in \mathcal{U}_0$ such that, with $z = (x, V(t, x))$, we have $Y_T^{t,z,\hat{\nu}} \geq g(X_T^{t,x,\hat{\nu}})$, $\mathbb{P}$–a.s. Then it follows from the geometric dynamic programming of Theorem 8.2 that $\mathbb{P}$–a.s:

$$Y_{t+h}^{t,z,\nu} = v(t,x) + \int_t^{t+h} b\left(s, Z_s^{t,z,\hat{\nu}}, \hat{\nu}_s\right) \mathrm{d}s + \int_t^{t+h} \hat{\nu}_s \cdot \mathrm{d}W_s \geq V\left(t+h, X_{t+h}^{t,x,\hat{\nu}}\right).$$

By Itô's formula, this implies that

$$\begin{aligned} 0 \leq \int_t^{t+h} &\left\{-\partial_t V(s, X_s^{t,x,\hat{\nu}})\right. \\ &\left. + H\left(s, Z_s^{t,z,\hat{\nu}}, DV(s, X_s^{t,x,\hat{\nu}}), D^2 V(s, X_s^{t,x,\hat{\nu}}), \hat{\nu}_s\right)\right\} \mathrm{d}s \\ &+ \int_t^{t+h} N^{\nu_s}\left(s, X_s^{t,x,\hat{\nu}}, DV(s, X_s^{t,x,\hat{\nu}})\right) \cdot \mathrm{d}W_s, \end{aligned} \tag{8.9}$$

where we introduced the functions:

$$H(t,x,y,p,A,u) := b(t,x,y,u) - \mu(t,x,u) \cdot p - \frac{1}{2}\operatorname{Tr}\left[\sigma(t,x,u)^2 A\right], \tag{8.10}$$

$$N^u(t,x,p) := u - \sigma(t,x,u)p. \tag{8.11}$$

We continue our intuitive derivation of the dynamic programming equation by assuming that all terms inside the integrals are bounded (we know that this can be achieved by localization). Then the first integral behaves like Ch, while the second integral can be viewed as a time-changed Brownian motion. By the properties of the Brownian motion, it follows that the integrand of the stochastic integral term must be zero at the origin:

$$N_t^{\nu_t}(t,x,DV(t,x)) = 0 \text{ or, equivalently, } \nu_t = \psi\big(t,x,DV(t,x)\big),$$

where ψ was introduced in (8.1). In particular, this implies that

$$\psi(t,x,DV(t,x)) \in U,$$

or, equivalently,

$$\delta_U(\xi) - \xi \cdot \psi(t,x,DV(t,x)) \geq 0 \quad \text{for all } \xi \in \tilde{U}_1, \tag{8.12}$$

by (8.7). Taking expected values in (8.9), normalizing by h, and sending h to zero, we see that

$$-\partial_t V(t,x) + H\big(t,x,V(t,x),DV(t,x),D^2V(t,x),\psi(t,x,DV(t,x))\big) \geq 0. \tag{8.13}$$

Putting (8.12) and (8.13) together, we obtain

$$\min\left\{-\partial_t V + H\big(.,V,DV,D^2V,\psi(.,DV)\big),\ \inf_{\xi\in\tilde{U}_1}\big(\delta_U(\xi) - \xi\cdot\psi(.,DV)\big)\right\} \geq 0.$$

By using the second part of the geometric dynamic programming principle, see Remark 8.3, we expect to prove that equality holds in the latter dynamic programming equation.

The Dynamic Programming Equation We next turn to a rigorous derivation of the dynamic programming equation. In the subsequent proof, we shall use the first part of the dynamic programming reported in Theorem 8.2 to prove that the stochastic target problem is a supersolution of the corresponding dynamic programming equation. For completeness, we will also provide the proof of the subsolution property based on the full dynamic programming principle of Remark 8.3. We observe however that our subsequent applications will only make use of the supersolution property.

Theorem 8.5. *Assume that V is locally bounded, and let the maps H and ψ be continuous. Then V is a viscosity supersolution of the dynamic programming equation on* **S**:

$$\min\left\{-\partial_t V + H\big(.,V,DV,D^2V,\psi(.,DV)\big),\ \inf_{\xi\in\tilde{U}_1}\big(\delta_U(\xi) - \xi\cdot\psi(.,DV)\big)\right\} = 0.$$

Assume further that ψ is locally Lipschitz continuous and U has nonempty interior. Then V is a viscosity solution of the above DPE on $\mathbf{S}$.

Proof. As usual, we prove separately the supersolution and the subsolution properties.

1. *Supersolution*: Let $(t_0, x_0) \in \mathbf{S}$ and $\varphi \in C^2(\mathbf{S})$ be such that

$$\text{(strict)} \min_{\mathbf{S}}(V_* - \varphi) = (V_* - \varphi)(t_0, x_0) = 0,$$

and assume to the contrary that

$$-2\eta := \left(-\partial_t V + H\left(., V, DV, D^2 V, \psi(., DV)\right)\right)(t_0, x_0) < 0. \quad (8.14)$$

(i) By the continuity of H and ψ, we may find $\varepsilon > 0$ such that

$$-\partial_t V(t,x) + H\left(t, x, y, DV(t,x), D^2 V(t,x), u\right) \leq -\eta$$
for $(t,x) \in B_\varepsilon(t_0, x_0)$, $|y - \varphi(t,x)| \leq \varepsilon$, and $u \in U$ s.t. $|N^u(t,x,p)| \leq \varepsilon$. (8.15)

Notice that (8.15) is obviously true if $\{u \in U : |N^u(t,x,p)| \leq \varepsilon\} = \emptyset$, so that the subsequent argument holds in this case as well.

Since (t_0, x_0) is a strict minimizer of the difference $V_* - \varphi$, we have

$$\gamma := \min_{\partial B_\varepsilon(t_0,x_0)} (V_* - \varphi) > 0. \quad (8.16)$$

(ii) Let $(t_n, x_n)_n \subset B_\varepsilon(t_0, x_0)$ be a sequence such that

$$(t_n, x_n) \longrightarrow (t_0, x_0) \text{ and } V(t_n, x_n) \longrightarrow V_*(t_0, x_0), \quad (8.17)$$

and set $y_n := V(t_n, x_n) + n^{-1}$ and $z_n := (x_n, y_n)$. By the definition of the problem $V(t_n, x_n)$, there exists a control process $\hat{\nu}^n \in \mathcal{U}_0$ such that the process $Z^n := Z^{t_n, z_n, \hat{\nu}^n}$ satisfies $Y_T^n \geq g(X_T^n)$, $\mathbb{P}$−a.s. Consider the stopping times:

$$\theta_n^0 := \inf\{t > t_n : (t, X_t^n) \notin B_\varepsilon(t_0, x_0)\},$$
$$\theta_n := \theta_n^0 \wedge \inf\{t > t_n : |Y_t^n - \varphi(t, X_t^n)| \geq \varepsilon\}.$$

Then, it follows from the geometric dynamic programming principle that

$$Y_{t \wedge \theta_n}^n \geq V\left(t \wedge \theta_n, X_{t \wedge \theta_n}^n\right).$$

Since $V \geq V_* \geq \varphi$, and using (8.16) and the definition of θ_n, this implies that

$$Y^n_{t\wedge\theta_n} \geq \varphi\left(t\wedge\theta_n, X^n_{t\wedge\theta_n}\right) + \mathbf{1}_{\{t=\theta_n\}}\left(\gamma\mathbf{1}_{\{\theta_n=\theta_n^o\}} + \varepsilon\mathbf{1}_{\{\theta_n<\theta_n^o\}}\right)$$
$$\geq \varphi\left(t\wedge\theta_n, X^n_{t\wedge\theta_n}\right) + (\gamma\wedge\varepsilon)\mathbf{1}_{\{t=\theta_n\}}. \tag{8.18}$$

(iii) Denoting $c_n := V(t_n, x_n) - \varphi(t_n, x_n) - n^{-1}$, we write the process Y^n as

$$Y^n_t = c_n + \varphi(t_n, x_n) + \int_{t_n}^t b(s, Z^n_s, \hat{\nu}^n_s)\mathrm{d}s + \int_{t_n}^t \hat{\nu}^n_s \cdot \mathrm{d}W_s.$$

Plugging this into (8.18) and applying Itô's formula, we then see that:

$$(\varepsilon\wedge\gamma)\mathbf{1}_{\{t=\theta_n\}} \leq c_n + \int_{t_n}^{t\wedge\theta_n} \delta^n_s \mathrm{d}s + \int_{t_n}^{t\wedge\theta_n} N^{\hat{\nu}^n_s}\left(s, X^n_s, D\varphi(s, X^n_s)\right)\cdot \mathrm{d}W_s$$
$$\leq M_n := c_n + \int_{t_n}^{t\wedge\theta_n} \delta^n_s \mathbf{1}_{A^n}(s)\mathrm{d}s + \int_{t_n}^{t\wedge\theta_n} N^{\hat{\nu}^n_s}\left(s, X^n_s, D\varphi(s, X^n_s)\right)\cdot \mathrm{d}W_s, \tag{8.19}$$

where

$$\delta^n_s := -\partial_t\varphi(s, X^n_s) + H\left(s, Z^n_s, D\varphi(s, X^n_s), D^2\varphi(s, X^n_s), \hat{\nu}_s\right)$$

and

$$A^n := \{s \in [t_n, \theta_n] : \delta^n_s > -\eta\}.$$

By (8.15), observe that the diffusion term $\zeta^n_s := N^{\hat{\nu}^n_s}\left(s, X^n_s, D\varphi(s, X^n_s)\right)$ in (8.19) satisfies $|\zeta^n_s| \geq \eta$ for all $s \in A^n$. Then, by introducing the exponential local martingale L^n defined by

$$L^n_{t_n} = 1 \text{ and } \mathrm{d}L^n_t = L^n_t|\zeta^n_t|^{-2}\zeta^n_t \cdot \mathrm{d}W_t, \quad t \geq t_n,$$

we see that the process $M^n L^n$ is a positive local martingale. Then $M^n L^n$ is a supermartingale, and it follows from (8.19) that

$$\varepsilon\wedge\gamma \leq \mathbb{E}\left[M^n_{\theta_n} L^n_{\theta_n}\right] \leq M^n_{t_n} L^n_{t_n} = c_n,$$

which can not happen because $c_n \longrightarrow 0$. Hence, our starting point (8.14) can not happen, and the proof of the supersolution property is complete.

2. *Subsolution*: Let $(t_0, x_0) \in \mathbf{S}$ and $\varphi \in C^2(\mathbf{S})$ be such that

$$\text{(strict)} \max_{\mathbf{S}}(V^* - \varphi) = (V^* - \varphi)(t_0, x_0) \;=\; 0, \tag{8.20}$$

and assume to the contrary that

$$\begin{aligned} 2\eta := \left(-\partial_t\varphi + H\left(., \varphi, D\varphi, D^2\varphi, \psi(., \varphi)\right)\right)(t_0, x_0) > 0, \\ \text{and } \inf_{\xi\in\tilde{U}_1} \left(\delta_U(\xi) - \xi\cdot\psi(., D\varphi)\right)(t_0, x_0) > 0. \end{aligned} \tag{8.21}$$

(i) By the continuity of H and ψ, and the characterization of $\operatorname{int}(U)$ in (8.8), it follows from (8.21) that

$$\begin{aligned} \left(-\partial_t\varphi + H\left(., y, D\varphi, D^2\varphi, \psi(., D\varphi)\right)\right) \geq \eta \text{ and } \psi(., D\varphi) \in U \\ \text{for } (t, x) \in B_\varepsilon(t_0, x_0) \text{ and } |y - \varphi(t, x)| \leq \varepsilon. \end{aligned} \tag{8.22}$$

Also, since (t_0, x_0) is a strict maximizer in (8.20), we have

$$-\zeta := \max_{\partial_p B_\varepsilon(t_0,x_0)} (V^* - \varphi) \;<\; 0, \tag{8.23}$$

where $\partial_p B_\varepsilon(t_0, x_0) \;:=\; \{t_0 + \varepsilon\} \times \operatorname{cl}\left(B_\varepsilon(t_0, x_0)\right) \cup \left[t_0, t_0 + \varepsilon\right) \times \partial B_\varepsilon(x_0)$ denotes the parabolic boundary of $B_\varepsilon(t_0, x_0)$.

(ii) Let $(t_n, x_n)_n$ be a sequence in $\mathbf{S}$ which converges to (t_0, x_0) and such that $V(t_n, x_n) \to V^*(t_0, x_0)$. Set $y_n = V(t_n, x_n) - n^{-1}$ and observe that

$$\gamma_n \;:=\; y_n - \varphi(t_n, x_n) \longrightarrow 0. \tag{8.24}$$

Let $Z^n \;:=\; (X^n, Y^n)$ denote the controlled state process associated to the Markovian control $\hat{\nu}^n_t \;=\; \psi(t, X^n_t, D\varphi(t, X^n_t))$ and the initial condition $Z^n_{t_n} \;=\; (x_n, y_n)$. Since ψ is locally Lipschitz-continuous, the process Z^n is well defined. We next define the stopping times

$$\begin{aligned} \theta^0_n &:= \inf\left\{s \geq t_n \;:\; (s, X^n_s) \notin B_\varepsilon(t_0, x_0)\right\}, \\ \theta_n &:= \theta^0_n \wedge \inf\left\{s \geq t_n \;:\; |Y^n(s) - \varphi(s, X^n_s)| \geq \varepsilon\right\}. \end{aligned}$$

By the first line in (8.22), (8.24) and a standard comparison theorem, it follows that $Y^n_{\theta_n} - \varphi(\theta_n, X^n_{\theta_n}) \geq \varepsilon$ on $\{|Y^n_{\theta_n} - \varphi(\theta_n, X^n_{\theta_n})| \geq \varepsilon\}$ for n large enough. Since $V \leq V^* \leq \varphi$, we then deduce from (8.23) and the definition of θ_n that

$$\begin{aligned} Y^n_{\theta_n} - V\left(\theta_n, X^n_{\theta^n}\right) \geq \mathbf{1}_{\{\theta_n<\theta^0_n\}}\left(Y^n_{\theta_n} - \varphi\left(\theta_n, X^n_{\theta_n}\right)\right) \\ +\mathbf{1}_{\{\theta_n=\theta^0_n\}}\left(Y^n_{\theta^0_n} - V^*\left(\theta^0_n, X^n_{\theta^0_n}\right)\right) \end{aligned}$$

$$\geq \varepsilon \mathbf{1}_{\{\theta_n < \theta_n^o\}} + \mathbf{1}_{\{\theta_n = \theta_n^o\}} \left(Y_{\theta_n^0}^n - V^*\left(\theta_n^0, X_{\theta_n^0}^n\right) \right)$$

$$\geq \varepsilon \mathbf{1}_{\{\theta_n < \theta_n^0\}} + \mathbf{1}_{\{\theta_n = \theta_n^0\}} \left(Y_{\theta_n^0}^n + \zeta - \varphi\left(\theta_n^0, X_{\theta_n^0}^n\right) \right)$$

$$\geq \varepsilon \wedge \zeta + \mathbf{1}_{\{\theta_n = \theta_n^0\}} \left(Y_{\theta_n^0}^n - \varphi\left(\theta_n^0, X_{\theta_n^0}^n\right) \right).$$

We continue by using Itô's formula:

$$Y_{\theta_n}^n - V\left(\theta_n, X_{\theta^n}^n\right) \geq \varepsilon \wedge \zeta + \mathbf{1}_{\{\theta_n = \theta_n^0\}} \left(\gamma_n + \int_{t_n}^{\theta_n} \alpha(s, X_s^n, Y_s^n) \mathrm{d}s \right),$$

where the drift term $\alpha(\cdot) \geq \eta$ is defined in (8.22) and the diffusion coefficient vanishes by the definition of the function ψ in (8.1). Since $\varepsilon, \zeta > 0$, and $\gamma_n \to 0$, this implies that

$$Y_{\theta_n}^n \;\geq\; V\left(\theta_n, X_{\theta^n}^n\right) \text{ for sufficiently large } n.$$

Recalling that the initial position of the process Y^n is $y_n = V(t_n, x_n) - n^{-1} < V(t_n, x_n)$, this is clearly in contradiction with the second part of the geometric dynamic programming principle discussed in Remark 8.3. □

8.1.4 Application: Hedging Under Portfolio Constraints

As an application of the previous results, we now study the problem of superhedging under portfolio constraints in the context of the Black–Scholes model.

Formulation We consider a financial market consisting of $d + 1$ assets. The first asset X^0 is nonrisky and is normalized to unity. The d next assets are risky with price process $X = (X^1, \ldots, X^d)^{\mathrm{T}}$ defined by the Black–Scholes model:

$$\mathrm{d}X_t = X_t \star \sigma \mathrm{d}W_t,$$

where σ is a constant symmetric nondegenerate matrix in $\mathbb{R}^d$, and $x \star \sigma$ is the square matrix in $\mathbb{R}^d$ with entries $(x \star \sigma)_{i,j} = x_i \sigma_{i,j}$.

Remark 8.6. We observe that the normalization of the first asset to unity does not entail any loss of generality as we can always reduce to this case by discounting or, in other words, by taking the price process of this asset as a numéraire.

Also, the formulation of the above process X as a martingale is not a restriction as our subsequent superhedging problem only involves the underlying probability measure through the corresponding zero-measure sets. Therefore, under the no-arbitrage condition (or more precisely, no free lunch with vanishing risk), we can reduce the model to the above martingale case by an equivalent change of measure.

Under the self-financing condition, the liquidation value of the portfolio is defined by the controlled state process:

$$\mathrm{d}Y_t^\pi = \sigma\pi_t \cdot \mathrm{d}W_t,$$

where π is the control process, with π_t^i representing the amount invested in the ith risky asset X^i at time t.

We introduce portfolio constraints by imposing that the portfolio process π must be valued in a subset U of $\mathbb{R}^d$. We shall assume that

$$U \text{ is closed convex subset of } \mathbb{R}^d,\ \mathrm{int}(U) \neq \emptyset, \text{ and } 0 \in U. \tag{8.25}$$

We then define the controls set by $\mathcal{U}_o$ as in the previous sections, and we define the superhedging problem under portfolio constraints by the stochastic target problem:

$$V(t,x) := \inf\left\{y : Y_T^{t,y,\pi} \geq g(X_T^{t,x}),\ \mathbb{P}-\text{a.s. for some } \pi \in \mathcal{U}_0\right\}, \tag{8.26}$$

where $g : \mathbb{R}_+^d \longrightarrow \mathbb{R}_+$ is a nonnegative LSC function with linear growth.

We shall provide an explicit solution of this problem by only using the supersolution claim from Theorem 8.5. This will provide a minorant of the superhedging cost V. To prove that this minorant is indeed the desired value function, we will use a verification argument.

Deriving a Minorant of the Superhedging Cost First, since $0 \leq g(x) \leq C(1 + |x|)$ for some constant $C > 0$, we deduce that $0 \leq V \leq C(1 + |x|)$, the right hand side inequality is easily justified by the buy-and-hold strategy suggested by the linear upper bound. Then, by a direct application of the first part of Theorem 8.5, we know that the LSC envelope V_* of V is a supersolution of the DPE:

$$-\partial_t V_* - \frac{1}{2}\mathrm{Tr}\left[(x \star \sigma)^2 D^2 V_*\right] \geq 0, \tag{8.27}$$

$$\delta_U(\xi) - \xi \cdot (x \star DV_*)) \geq 0, \quad \text{for all } \xi \in \tilde{U}. \tag{8.28}$$

Notice that (8.28) is equivalent to

$$\text{the map } \lambda \longmapsto h(\lambda) := \lambda\delta_U(\xi) - V_*(t, x \star \mathrm{e}^{\lambda\xi}) \text{ is that nondecreasing}, \tag{8.29}$$

where $\mathrm{e}^{\lambda\xi}$ is the vector of $\mathbb{R}^d$ with entries $(\mathrm{e}^{\lambda\xi})_i = \mathrm{e}^{\lambda\xi_i}$. Then $h(1) \geq h(0)$ provides

$$V_*(t,x) \geq \sup_{\xi\in\tilde{U}} V_*\left(x \star \mathrm{e}^{\xi}\right) - \delta_U(\xi).$$

We next observe that $V_*(T,.) \geq g$ (just use the definition of V, and send $t \nearrow T$). Then, we deduce from the previous inequality that

$$V_*(T,x) \geq \hat{g}(x) := \sup_{\xi\in\tilde{U}} g\left(x \star \mathrm{e}^{\xi}\right) - \delta_U(\xi) \quad \text{for all } x \in \mathbb{R}_+^d. \tag{8.30}$$

In other words, in order to superhedge the derivative security with final payoff $g(X_T)$, the constraints on the portfolio require that one hedges the derivative security with larger payoff $\hat{g}(X_T)$. The function $\hat{g}$ is called the *face-lifted* payoff, and is the smallest majorant of g which satisfies the gradient constraint $x \star Dg(x) \in U$ for all $x \in \mathbb{R}^d_+$.

Combining (8.30) with (8.27), it follows from the comparison result for the linear Black–Scholes PDE that

$$V(t,x) \geq V_*(t,x) \geq v(t,x) := \mathbb{E}\left[\hat{g}(X_T^{t,x})\right] \quad \text{for all } (t,x) \in \mathbf{S}. \tag{8.31}$$

Explicit Solution Our objective is now to prove that $V = v$. To see this, consider the Black–Scholes hedging strategy $\hat{\pi}$ of the derivative security $\hat{g}(X_T^{t,x})$:

$$v(t,x) + \int_t^T \hat{\pi}_s \cdot \sigma \mathrm{d}W_s = \hat{g}(X_T^{t,x}).$$

Since $\hat{g}$ has linear growth, it follows that $\hat{\pi} \in \mathbb{H}^2$. We also observe that the random variable $\ln X_T^{t,x}$ is gaussian, so that the function v can be written in

$$v(t,x) = \int \hat{g}(\mathrm{e}^w) \frac{1}{\sqrt{2\pi\sigma^2(T-t)}} \mathrm{e}^{-\frac{1}{2}\left(\frac{w-x+\frac{1}{2}\sigma^2(T-t)}{\sigma\sqrt{T-t}}\right)^2} \mathrm{d}w.$$

Under this form, it is clear that v is a smooth function. Then the above hedging portfolio is given by

$$\hat{\pi}_s := X_s^{t,x} \star DV(s, X_s^{t,x}).$$

Notice that, for all $\xi \in \tilde{U}$,

$$\lambda\delta_U(\xi) - v(t, x\mathrm{e}^{\lambda\xi}) = \mathbb{E}\left[\lambda\delta_U(\xi) - \hat{g}\left(X_T^{t,x\mathrm{e}^{\lambda\xi}}\right)\right]$$

is nondecreasing in λ by applying (8.29) to $\hat{g}$ which, by definition, satisfies $x \star Dg(x) \in U$ for all $x \in \mathbb{R}^d_+$. Then, $x \star Dg(x) \in U$, and therefore the above replicating portfolio $\hat{\pi}$ takes values in U. Since $\hat{g} \geq g$, we deduce from (8.30) that $v \geq V$.

8.2 Stochastic Target Problem with Controlled Probability of Success

In this section, we extend the above problem to the case where the target has to be reached only with a given probability p:

$$\hat{V}(t,x,p) := \inf\left\{y \in \mathbb{R}_+ : \mathbb{P}\left[Y_T^{t,x,y,\nu} \geq g(X_T^{t,x,\nu})\right] \geq p \text{ for some } \nu \in \mathcal{U}_0\right\}. \tag{8.32}$$

In order to avoid degenerate results, we restrict the analysis to the case where the Y process takes non-negative values, by simply imposing the following conditions on the coefficients driving its dynamics:

$$b(t,x,0,u) \geq 0 \quad \text{for all}(t,x) \in \mathbf{S}, u \in U. \tag{8.33}$$

Notice that the above definition implies that

$$0 \; = \; \hat{V}(.,0) \; \leq \; \hat{V} \; \leq \; \hat{V}(.,1) \; = \; V, \tag{8.34}$$

and

$$\hat{V}(.,p) \;=\; 0 \text{ for } p < 0 \text{ and } \hat{V}(.,p) = \infty \;\; \text{for } p > 1\;, \tag{8.35}$$

with the usual convention $\inf \emptyset = \infty$.

8.2.1 Reduction to a Stochastic Target Problem

Our first objective is to convert this problem into a (standard) stochastic target problem, so as to apply the geometric dynamic programming arguments of the previous section.

To do this, we introduce an additional controlled state variable:

$$P_s^{t,p,\alpha} := p + \int_t^s \alpha_r \cdot \mathrm{d}W_r, \quad \text{for } s \in [t,T], \tag{8.36}$$

where the additional control α is an $\mathbb{F}$–progressively measurable $\mathbb{R}^d$–valued process satisfying the integrability condition $\mathbb{E}\left[\int_0^T |\alpha_s|^2 \mathrm{d}s\right] < \infty$. We then set $\hat{X} := (X,P)$, $\hat{\mathbf{S}} := [0,T) \times \mathbb{R}^d \times (0,1)$, $\hat{U} := U \times \mathbb{R}^d$, and denote by $\hat{\mathcal{U}}$ the corresponding set of admissible controls. Finally, we introduce the function:

$$G(\hat{x},y) := \mathbf{1}_{\{y \geq g(x)\}} - p \quad \text{for } y \in \mathbb{R},\; \hat{x} := (x,p) \in \mathbb{R}^d \times [0,1].$$

Proposition 8.7. *For all $t \in [0,T]$ and $\hat{x} = (x,p) \in \mathbb{R}^d \times [0,1]$, we have*

$$\hat{V}(t,\hat{x}) = \inf\left\{y \in \mathbb{R}_+ \;:\; G\left(\hat{X}_T^{t,\hat{x},\hat{\nu}}, Y_T^{t,x,y,\nu}\right) \geq 0 \textit{ for some } \hat{\nu} = (\nu,\alpha) \in \hat{\mathcal{U}}\right\}.$$

Proof. We denote by $v(t,x,p)$ the value function appearing on the right hand.

We first show that $\hat{V} \geq v$. For $y > \hat{V}(t,x,p)$, we can find $\nu \in \mathcal{U}$ such that $p_0 := \mathbb{P}\left[Y_T^{t,x,y,\nu} \geq g\big(X_T^{t,x,\nu}\big)\right] \geq p$. By the stochastic integral representation theorem, there exists an $\mathbb{F}$-progressively measurable process α such that

$$\mathbf{1}_{\{Y_T^{t,x,y,\nu} \geq g(X_T^{t,x,\nu})\}} = p_0 + \int_t^T \alpha_s \cdot \mathrm{d}W_s = P_T^{t,p_0,\alpha} \text{ and } \mathbb{E}\left[\int_t^T |\alpha_s|^2 \mathrm{d}s\right] < \infty.$$

Since $p_0 \geq p$, it follows that $\mathbf{1}_{\{Y_T^{t,x,y,\nu} \geq g(X_T^{t,x,\nu})\}} \geq P_T^{t,p,\alpha}$, and therefore $y \geq v(t,x,p)$ from the definition of the problem v.

We next show that $v \geq \hat{V}$. For $y > v(t,x,p)$, we have $G\left(\hat{X}_T^{t,\hat{x},\hat{\nu}}, Y_T^{t,x,y,\nu}\right) \geq 0$ for some $\hat{\nu} = (\nu, \alpha) \in \hat{\mathcal{U}}$. Since $P_{t,p}^{\alpha}$ is a martingale, it follows that

$$\mathbb{P}\left[Y_T^{t,x,y,\nu} \geq g(X_T^{t,x,\nu})\right] = \mathbb{E}\left[\mathbf{1}_{\{Y_T^{t,x,y,\nu} \geq g(X_T^{t,x,\nu})\}}\right] \geq \mathbb{E}\left[P_T^{t,p,\alpha}\right] = p,$$

which implies that $y \geq \hat{V}(t,x,p)$ by the definition of $\hat{V}$. □

Remark 8.8. 1. Suppose that the infimum in the definition of $\hat{V}(t,x,p)$ is achieved and there exists a control $\nu \in \mathcal{U}_0$ satisfying $\mathbb{P}\left[Y_T^{t,x,y,\nu} \geq g\left(X_T^{t,x,\nu}\right)\right] = p$, the above argument shows that:

$$P_s^{t,p,\alpha} = \mathbb{P}\left[Y_T^{t,x,y,\nu} \geq g\left(X_T^{t,x,\nu}\right) \middle| \mathcal{F}_s\right] \quad \text{for all } s \in [t,T].$$

2. It is easy to show that one can moreover restrict to control α such that the process $P^{t,p,\alpha}$ takes values in $[0,1]$. This is rather natural since this process should be interpreted as a conditional probability, and this corresponds to the natural domain $[0,1]$ of the variable p. We shall however avoid to introduce this state constraint, and use the fact that the value function $\hat{V}(\cdot,p)$ is constant for $p \leq 0$ and equal ∞ for $p > 1$, see (8.35).

8.2.2 The Dynamic Programming Equation

The above reduction of the problem $\hat{V}$ to a stochastic target problem allows to apply the geometric dynamic programming principle of the previous section, and to derive the corresponding dynamic programming equation. For $\hat{u} = (u,\alpha) \in \hat{U}$ and $\hat{x} = (x,p) \in \mathbb{R}^d \times [0,1]$, set

$$\hat{\mu}(\hat{x},\hat{u}) := \begin{pmatrix} \mu(x,u) \\ 0 \end{pmatrix}, \quad \hat{\sigma}(\hat{x},\hat{u}) := \begin{pmatrix} \sigma(x,u) \\ \alpha^{\mathrm{T}} \end{pmatrix}.$$

For $(y,q,A) \in \mathbb{R} \times \mathbb{R}^{d+1} \times \mathcal{S}_{d+1}$ and $\hat{u} = (u,\alpha) \in \hat{U}$,

$$\hat{N}^{\hat{u}}(t,\hat{x},y,q) := u - \hat{\sigma}(t,\hat{x},\hat{u})q = N^u(t,x,q_x) - q_p\alpha \text{ for } q = (q_x,q_p) \in \mathbb{R}^d \times \mathbb{R},$$

and we assume that

$$u \longmapsto N^u(t,x,q_x) \text{ is one to one, with inverse function } \psi(t,x,q_x). \tag{8.37}$$

Then, by a slight extension of Theorem 8.5, the corresponding dynamic programming equation is given by

$$0 = -\partial_t \hat{V} + \sup_{\alpha} \Big\{ b(., \hat{V}, \psi(., D_x \hat{V}, \alpha D_p \hat{V})) - \mu(., \psi(., D_x \hat{V}, \alpha D_p \hat{V})).D_x \hat{V}$$

$$-\frac{1}{2}\mathrm{Tr}\left[\sigma(., \psi(., D_x \hat{V}, \alpha D_p \hat{V}))^2 D_x^2 \hat{V}\right]$$

$$-\frac{1}{2}\alpha^2 D_p^2 \hat{V} - \alpha\sigma(., \psi(., D_x \hat{V}, \alpha D_p \hat{V})) D_{xp} \hat{V} \Big\}.$$

The application in the subsequent section will be only making use of the supersolution property of the stochastic target problem.

8.2.3 Application: Quantile Hedging in the Black–Scholes Model

The problem of quantile hedging was solved by Föllmer and Leukert [21] in the general model of asset prices process (non-necessarily Markovian), by means of the Neyman–Pearson lemma from mathematical statistics. The stochastic control approach developed in the present section allows to solve this type of problems in a wider generality. The objective of this section is to recover the explicit solution of [21] in the context of a complete financial market where the underlying risky asset prices are not affected by the control:

$$\mu(x,u) = \mu(x) \text{ and } \sigma(x,u) = \sigma(x) \text{ are independent of } u, \tag{8.38}$$

where μ and σ are Lipschitz-continuous, and $\sigma(x)$ is invertible for all x.

Notice that we will be only using the supersolution property from the results of the previous sections.

The Financial Market The process X, representing the price process of d risky assets, is defined by $X_t^{t,x} = x \in (0,\infty)^d$, and

$$\mathrm{d}X_s^{t,x} = X_s^{t,x} \star \sigma(X_s^{t,x}) \left(\lambda(X_s^{t,x})\mathrm{d}s + \mathrm{d}W_s\right) , where\ \lambda := \sigma^{-1}\mu.$$

We assume that the coefficients μ and σ are such that $X^{t,x} \in (0,\infty)^d$ $\mathbb{P}$–a.s. for all initial conditions $(t,x) \in [0,T] \times (0,\infty)^d$. In order to avoid arbitrage, we also assume that σ is invertible and that

$$\sup_{x \in (0,\infty)^d} |\lambda(x)| < \infty. \tag{8.39}$$

The drift coefficient of the controlled process Y is given by:

$$b(t,x,y,u) = u \cdot \lambda(x). \tag{8.40}$$

The control process ν is valued in $U = \mathbb{R}^d$, with components ν_s^i indicating the dollar investment in the i-th security at time s. After the usual reduction of the interest rates to zero, it follows from the self-financing condition that the liquidation value of the portfolio is given by

$$Y_s^{t,x,y,\nu} = y + \int_t^s \nu_r \cdot \sigma(X_s^{t,x}) \left(\lambda(X_s^{t,x})\mathrm{d}s + \mathrm{d}W_s\right), \quad s \geq t ,$$

The Quantile Hedging Problem The quantile hedging problem of the derivative security $g(X_T^{t,x})$ is defined by the stochastic target problem with controlled probability of success:

$$\hat{V}(t,x,p) := \inf\left\{y \in \mathbb{R}_+ \, : \, \mathbb{P}\left[Y_T^{t,x,y,\nu} \geq g(X_T^{t,x})\right] \geq p \text{ for some } \nu \in \mathcal{U}_0\right\}.$$

We shall assume throughout that $0 \leq g(x) \leq C(1+|x|)$ for all $x \in \mathbb{R}_+^d$. By the usual buy-and-hold hedging strategies, this implies that $0 \leq V(t,x) \leq C(1+|x|)$.

Under the above assumptions, the corresponding super hedging cost $V(t,x) := \hat{V}(t,x,1)$ is continuous and is given by

$$V(t,x) = \mathbb{E}^{\mathbb{Q}^{t,x}}\left[g(X_T^{t,x})\right],$$

where $\mathbb{Q}^{t,x}$ is the $\mathbb{P}$-equivalent martingale measure defined by

$$\frac{\mathrm{d}\mathbb{Q}^{t,x}}{\mathrm{d}\mathbb{P}} = \exp\left(-\frac{1}{2}\int_t^T |\lambda(X_s^{t,x})|^2 \mathrm{d}s - \int_t^T \lambda(X_s^{t,x}) \cdot \mathrm{d}W_s\right).$$

In particular, V is a viscosity solution on $[0,T) \times (0,\infty)^d$ of the linear PDE:

$$0 = -\partial_t V - \frac{1}{2}\mathrm{Tr}\left[(x \star \sigma)^2 D_x^2 V\right]. \tag{8.41}$$

For later use, let us denote by

$$W^{\mathbb{Q}^{t,x}} := W + \int_t^{\cdot} \lambda(X_s^{t,x})\mathrm{d}s, \; s \in [t,T],$$

the $\mathbb{Q}^{t,x}$-Brownian motion defined on $[t,T]$.

The Viscosity Supersolution Property By the results of the previous section, we have $\hat{V}_*$ is a viscosity supersolution on $[0,T) \times \mathbb{R}_+^d \times [0,1]$ of the equation

$$0 \le -\partial_t \hat{V}_* - \frac{1}{2}\mathrm{Tr}\left[\sigma^2 x^2 D_x^2 \hat{V}_*\right]$$
$$- \inf_{\alpha \in \mathbb{R}^d} \left(-\alpha\lambda D_p \hat{V}_* + \mathrm{Tr}\left[\sigma x \alpha D_{xp} \hat{V}_*\right] + \frac{1}{2}|\alpha|^2 D_p^2 \hat{V}_*\right). \qquad (8.42)$$

The boundary conditions at $p = 0$ and $p = 1$ are immediate:

$$\hat{V}_*(\cdot, 1) = V \text{ and } \hat{V}_*(\cdot, 0) = 0 \text{ on } [0, T] \times \mathbb{R}_+^d. \qquad (8.43)$$

We next determine the boundary condition at the terminal time $t = T$.

Lemma 8.9. *For all $x \in \mathbb{R}_+^d$ and $p \in [0, 1]$, we have $\hat{V}_*(T, x, p) \ge pg(x)$.*

Proof. Let $(t_n, x_n, p_n)_n$ be a sequence in $[0, T) \times \mathbb{R}_+^d \times (0, 1)$ converging to (T, x, p) with $\hat{V}(t_n, x_n, p_n) \longrightarrow \hat{V}_*(T, x, p)$, and consider $y_n := \hat{V}(t_n, x_n, p_n) + 1/n$. By definition of the quantile hedging problem, there is a sequence $(\nu_n, \alpha_n) \in \hat{\mathcal{U}}_0$ such that

$$\mathbf{1}_{\{Y_T^{t_n,x_n,y_n,\nu_n} - g(X_T^{t_n,x_n}) \ge 0\}} \ge P_T^{t_n,p_n,\alpha_n}.$$

This implies that

$$Y_T^{t_n,x_n,y_n,\nu_n} \ge P_T^{t_n,p_n,\alpha_n} g(X_T^{t_n,x_n}).$$

Taking the expectation under $\mathbb{Q}^{t_n,x_n}$, this provides:

$$y_n \ge \mathbb{E}^{\mathbb{Q}^{t_n,x_n}}\left[Y_T^{t_n,x_n,y_n,\nu_n}\right] \ge \mathbb{E}^{\mathbb{Q}^{t_n,x_n}}\left[P_T^{t_n,p_n,\alpha_n} g(X_T^{t_n,x_n})\right]$$
$$= \mathbb{E}\left[L_T^{t_n,x_n} P_T^{t_n,p_n,\alpha_n} g(X_T^{t_n,x_n})\right],$$

where we denote $L_T^{t_n,x_n} := \exp\left(-\int_{t_n}^T \lambda(X_s^{t_n,x_n}) \cdot \mathrm{d}W_s - \frac{1}{2}\int_{t_n}^T |\lambda(X_s^{t_n,x_n})|^2 \mathrm{d}s\right)$. Then

$$y_n \ge \mathbb{E}\left[P_T^{t_n,p_n,\alpha_n} g(x)\right] + \mathbb{E}\left[P_T^{t_n,p_n,\alpha_n}\left(L_T^{t_n,x_n} g(X_T^{t_n,x_n}) - g(x)\right)\right]$$
$$= p_n g(x) + \mathbb{E}\left[P_T^{t_n,p_n,\alpha_n}\left(L_T^{t_n,x_n} g(X_T^{t_n,x_n}) - g(x)\right)\right]$$
$$\ge p_n g(x) - \mathbb{E}\left[P_T^{t_n,p_n,\alpha_n}\left|L_T^{t_n,x_n} g(X_T^{t_n,x_n}) - g(x)\right|\right], \qquad (8.44)$$

where we used the fact that P^{t_n,p_n,α_n} is a nonnegative martingale. Now, since this process is also bounded by 1, we have

$$\mathbb{E}\left[P_T^{t_n,p_n,\alpha_n}\left|L_T^{t_n,x_n} g(X_T^{t_n,x_n}) - g(x)\right|\right] \le \mathbb{E}\left[\left|L_T^{t_n,x_n} g(X_T^{t_n,x_n}) - g(x)\right|\right] \longrightarrow 0$$

as $n \to \infty$, by the stability properties of the flow and the dominated convergence theorem. Then, by taking limits in (8.44), we obtain that $\hat{V}_*(T,x,p) = \lim_{n\to\infty} y_n \geq pg(x)$, which is the required inequality. □

An Explicit Minorant of $\hat{V}$ The key idea is to introduce the Legendre-Fenchel dual of V_* with respect to the $p-$variable in order to remove the non-linearity in (8.42):

$$v(t,x,q) := \sup_{p\in\mathbb{R}} \left\{ pq - \hat{V}_*(t,x,p) \right\}, (t,x,q) \in [0,T] \times (0,\infty)^d \times \mathbb{R}. \quad (8.45)$$

By the definition of the function $\hat{V}$, we have

$$v(.,q) = \infty \text{ for } q < 0 \text{ and } v(.,q) = \sup_{p\in[0,1]} \left\{ pq - \hat{V}_*(.,p) \right\} \text{ for } q > 0. \quad (8.46)$$

Using the above supersolution property of $\hat{V}_*$, we shall prove below that v is an upper-semicontinuous viscosity subsolution on $[0,T) \times (0,\infty)^d \times (0,\infty)$ of

$$-\partial_t v - \frac{1}{2}\mathrm{Tr}\left[\sigma^2 x^2 D_x^2 v\right] - \frac{1}{2}|\lambda|^2 q^2 D_q^2 v - \mathrm{Tr}\left[\lambda\sigma x D_{xq} v\right] \leq 0 \quad (8.47)$$

with the boundary condition

$$v(T,x,q) \leq (q - g(x))^+ . \quad (8.48)$$

Since the above equation is linear, we deduce from the comparison result an explicit upper bound for v given by the Feynman-Kac representation result:

$$v(t,x,q) \leq \bar{v}(t,x,q) := \mathbb{E}^{\mathbb{Q}_{t,x}}\left[\left(Q_T^{t,x,q} - g(X_T^{t,x})\right)^+\right], \quad (8.49)$$

on $[0,T] \times (0,\infty)^d \times (0,\infty)$, where the process $Q^{t,x,q}$ is defined by the dynamics

$$\frac{\mathrm{d}Q_s^{t,x,q}}{Q_s^{t,x,q}} = \lambda(X_s^{t,x}) \cdot \mathrm{d}W_s^{\mathbb{Q}^{t,x}} \text{ and } Q^{t,x,q}(t) = q \in (0,\infty). \quad (8.50)$$

Given the explicit representation of $\bar{v}$, we can now provide a lower bound for the primal function $\hat{V}$ by using (8.46).

We next deduce from (8.49) a lower bound for the quantile hedging problem $\hat{V}$. Recall that the convex envelope $\hat{V}_*^{\mathrm{conv}\,p}$ of $\hat{V}_*$ with respect to p is given by the bi-conjugate function

$$\hat{V}_*^{\mathrm{conv}\,p}(t,x,p) = \sup_q \left\{ pq - v(t,x,q) \right\}$$

Chapter 9
Second Order Stochastic Target Problems

In this chapter, we extend the class of stochastic target problems of the previous section to the case where the quadratic variation of the control process ν is involved in the optimization problem. This new class of problems is motivated by applications in financial mathematics.

We first start by studying in details the so-called problem of hedging under gamma constraints. This simple example already outlines the main difficulties. By using the tools from viscosity solutions, we shall first exhibit a minorant for the superhedging cost in this context. We then argue by verification to prove that this minorant is indeed the desired value function.

We then turn to a general formulation of second-order stochastic target problems. Of course, in general, there is no hope to use a verification argument as in the example of the first section. We therefore provide the main tools in order to show that the value function is a viscosity solution of the corresponding dynamic programming equation.

Finally, Sect. 9.3 provides another application to the problem of superhedging under illiquidity cost. We will consider the illiquid financial market introduced by Cetin et al. [8], and we will show that our second-order stochastic target framework leads to an illiquidity cost which can be characterized by means of a nonlinear PDE.

9.1 Superhedging Under Gamma Constraints

In this section, we focus on an alternative constraint on the portfolio π. For simplicity, we consider a financial market with a single risky asset. Let $Z_t(\omega) := S_t^{-1}\pi_t(\omega)$ denote the vector of number of shares of the risky assets held at each time t and $\omega \in \Omega$. By definition of the portfolio strategy, the investor has to adjust his strategy at each time t, by passing the number of shares from Z_t to $Z_{t+\mathrm{d}t}$. His demand in risky assets at time t is then given by "$\mathrm{d}Z_t$".

N. Touzi, *Optimal Stochastic Control, Stochastic Target Problems, and Backward SDE*, 123
Fields Institute Monographs 29, DOI 10.1007/978-1-4614-4286-8_9,

In an equilibrium model, the price process of the risky asset would be pushed upward for a large demand of the investor. We therefore study the hedging problem with constrained portfolio adjustment.

However, market practitioners only focus on the diffusion component of the hedge adjustment dZ_t, which is given by the so-called *gamma*, i.e., the Hessian of the Black–Scholes prices. The gamma of the hedging strategy is an important risk control parameter indicating the size of the replacement of the hedging portfolio induced by a stress scenario, i.e., a sudden jump of the underlying asset spot price. A large portfolio gamma leads to two different risks depending on its sign:

- A large positive gamma requires that the seller of the option adjusts his hedging portfolio by a large purchase of the underlying asset. This is a typical risk that traders want to avoid because then the price to be paid for this hedging adjustment is very high and sometimes even impossible because of the limited offer of underlying assets on the financial market.
- A negative gamma induces a risk of different nature. Indeed, the hedger has the choice between two alternative strategies: either adjust the hedge at the expense of an outrageous market price, or hold the Delta position risk. The latter buy-and-hold strategy does not violate the hedge thanks to the (local) concavity of the payoff (negative gamma). There are two ways to understand this result: the second-order term in the Taylor expansion has a sign in favor of the hedger, or equivalently, the option price curve is below its tangent which represents the buy-and-hold position.

This problem turns out to present serious mathematical difficulties. The analysis of this section provides a solution of the problem of hedging under upper bound on the gamma in a very specific situation. The lower bound on the gamma introduces more difficulties due to the fact that the nonlinearity in the "first guess" equation is not elliptic.

9.1.1 Problem Formulation

We consider a financial market which consists of one bank account, with constant price process $S_t^0 = 1$ for all $t \in [0, T]$, and one risky asset with price process evolving according to the Black–Scholes model:

$$S_u := S_t \exp\left(-\frac{1}{2}\sigma^2(t-u) + \sigma(W_t - W_u)\right), \qquad t \leq u \leq T.$$

Here, W is a standard Brownian motion in $\mathbb{R}$ defined on a complete probability space $(\Omega, \mathcal{F}, P)$. We shall denote by $\mathbb{F} = \{\mathcal{F}_t, 0 \leq t \leq T\}$ the P-augmentation of the filtration generated by W.

Observe that there is no loss of generality in taking S as a martingale, as one can always reduce the model to this case by judicious change of measure. On the other hand, the subsequent analysis can be easily extended to the case of a varying volatility coefficient.

We denote by $Z = \{Z_u,\ t \le u \le T\}$ the process of number of shares of risky asset S held by the agent during the time interval $[t, T]$. Then, by the self-financing condition, the wealth process induced by some initial capital y, at time t, and portfolio strategy Z is given by:

$$Y_u = y + \int_t^u Z_r \mathrm{d}S_r, \qquad t \le u \le T.$$

In order to introduce constraints on the variations of the hedging portfolio Z, we restrict Z to the class of continuous semimartingales with respect to the filtration $\mathbb{F}$. Since $\mathbb{F}$ is the Brownian filtration, we define the controlled portfolio strategy $Z^{z,\alpha,\Gamma}$ by

$$Z_u^{z,\alpha,\Gamma} = z + \int_t^u \alpha_r \mathrm{d}r + \int_t^u \Gamma_r \sigma \mathrm{d}W_r, \quad t \le u \le T, \tag{9.1}$$

where $z \in \mathbb{R}$ is the time t initial portfolio and the *control* pair (α, Γ) are bounded progressively measurable processes. We denote by $\mathcal{B}_t$ the collection of all such control processes.

Hence, a *trading strategy* is defined by the triple $\nu := (z, \alpha, \Gamma)$ with $z \in \mathbb{R}$ and $(\alpha, \Gamma) \in \mathcal{B}_t$. The associated wealth process, denoted by $Y^{y,\nu}$, is given by

$$Y_u^{y,\nu} = y + \int_t^u Z_r^\nu \mathrm{d}S_r, \qquad t \le u \le T, \tag{9.2}$$

where y is the time t initial capital. We now formulate the gamma constraint in the following way. Let $\underline{\Gamma} < 0 < \overline{\Gamma}$ be two fixed constants. Given some initial capital $y \in \mathbb{R}$, we define the set of y*-admissible* trading strategies by

$$\mathcal{A}_t(\underline{\Gamma}, \overline{\Gamma}) := \left\{\nu = (y, \alpha, \Gamma) \in \mathbb{R} \times \mathcal{B}_t \ : \ \underline{\Gamma} \le \Gamma_\cdot \le \overline{\Gamma}\right\}.$$

As in the previous sections, we consider the super-replication problem of some European-type contingent claim $g(S_T)$

$$v(t, S_t) := \inf\left\{y \ : \ Y_T^{y,\nu} \ge g(S_T) \text{ a.s. for some } \nu \in \mathcal{A}_t(\underline{\Gamma}, \overline{\Gamma})\right\}. \tag{9.3}$$

Remark 9.1. The above set of admissible strategies seems to be very restrictive. We will see later that one can possibly enlarge, but not so much! The fundamental reason behind this can be understood from the following result due to Bank and Baum [1] and restated here in the case of the Brownian motion:

Lemma 9.2. *Let ϕ be a progressively measurable process with $\int_0^1 |\phi_t|^2 dt < \infty$, $\mathbb{P}$–a.s. Then for all $\varepsilon > 0$, there is a process ϕ^ε with $d\phi_t^\varepsilon = \alpha_t^\varepsilon dt$ for some progressively measurable α^ε with $\int_0^1 |\alpha_t^\varepsilon| dt < \infty$, $\mathbb{P}$–a.s. such that*

$$\sup_{t \le 1} \left\| \int_0^1 \phi_t dW_t - \int_0^1 \phi_t^\varepsilon dW_t \right\|_{\mathbb{L}^\infty} \le \varepsilon.$$

Given this result, it is clear that without any constraint on the process α in the strategy ν, the superhedging cost would be obviously equal to the Black–Scholes unconstrained price. Indeed, the previous lemma says that one can approximate the Black–Scholes hedging strategy by a sequence of hedging strategies with zero gamma without affecting the liquidation value of the hedging portfolio by more that ε.

9.1.2 Hedging Under Upper Gamma Constraint

In this section, we consider the case

$$\underline{\Gamma} = -\infty \text{ and we denote } \overline{\mathcal{A}}_t(\overline{\Gamma}) := \mathcal{A}_t(-\infty, \overline{\Gamma}).$$

Our goal is to derive the following explicit solution: $v(t, S_t)$ is the (unconstrained) Black–Scholes price of some convenient *face-lifted* contingent claim $\hat{g}(S_T)$, where the function $\hat{g}$ is defined by

$$\hat{g}(s) := h^{\text{conc}}(s) + \overline{\Gamma} s \ln s \text{ with } h(s) := g(s) - \overline{\Gamma} s \ln s,$$

and h^{conc} denotes the concave envelope of h. Observe that this function can be computed easily. The reason for introducing this function is the following:

Lemma 9.3. *$\hat{g}$ is the smallest function satisfying the conditions*

$$\text{(i) } \hat{g} \ge g \text{ , and (ii) } s \longmapsto \hat{g}(s) - \overline{\Gamma} s \ln s \text{ is concave.}$$

The proof of this easy result is omitted.

Theorem 9.4. *Let g be a lower-semicontinuous mapping on $\mathbb{R}_+$ with*

$$s \longmapsto \hat{g}(s) - C\, s \ln s \text{ convex for some constant } C. \tag{9.4}$$

Then the value function (9.3) is given by

$$v(t, s) = E_{t,s}\left[\hat{g}(S_T)\right] \quad \text{for all } (t, s) \in [0, T) \times (0, \infty).$$

Discussion 1. We first make some comments on the model. Intuitively, we expect the optimal hedging portfolio to satisfy

$$\hat{Z}_u = v_s(u, S_u),$$

where v is the minimal superhedging cost. Assuming enough regularity, it follows from Itô's formula that

$$\mathrm{d}\hat{Z}_u = A_u \mathrm{d}u + \sigma S_u v_{ss}(u, S_u)\mathrm{d}W_u,$$

where $A(u)$ is given in terms of derivatives of v. Compare this equation with (9.1) to conclude that the associated *gamma* is

$$\hat{\Gamma}_u = S_u\, v_{ss}(u, S_u).$$

Therefore, the bound on the process $\hat{\Gamma}$ translates to a bound on sv_{ss}. Notice that, by changing the definition of the process Γ in (9.1), we may bound v_{ss} instead of sv_{ss}. However, we choose to study sv_{ss} because it is a dimensionless quantity, i.e., if all the parameters in the problem are increased by the same factor, sv_{ss} still remains unchanged.

2. Intuitively, we expect to obtain a similar type solution to the case of portfolio constraints. If the Black–Scholes solution happens to satisfy the gamma constraint, then it solves the problem with gamma constraint. In this case v satisfies the PDE $-\partial_t v - \mathcal{L}v = 0$. Since the Black–Scholes solution does not satisfy the gamma constraint, in general, we expect that the function v solves the variational inequality

$$\min\left\{-\partial_t v - \mathcal{L}v, \overline{\Gamma} - sv_{ss}\right\} = 0. \tag{9.5}$$

3. An important feature of the log-normal Black and Scholes model is that the variational inequality (9.5) reduces to the Black–Scholes PDE $-\partial_t v - \mathcal{L}v = 0$ as long as the terminal condition satisfies the gamma constraint (in a weak sense). From Lemma 9.3, the *face-lifted* payoff function $\hat{g}$ is precisely the minimal function above g which satisfies the gamma constraint (in a weak sense). This explains the nature of the solution reported in Theorem 9.4, namely, $v(t, S_t)$ is the Black–Scholes price of the contingent claim $\hat{g}\,(S_T)$.

Dynamic Programming and Viscosity Property. We now turn to the proof of Theorem 9.4. We shall denote

$$\hat{v}(t, s) := E_{t,s}\left[\hat{g}\,(S_T)\right].$$

It is easy to check that $\hat{v}$ is a smooth function satisfying

$$\partial_t + \mathcal{L}\hat{v} \;=\; 0 \text{ and } s\hat{v}_{ss} \;\leq\; \overline{\Gamma} \text{ on } [0, T) \times (0, \infty)\,. \tag{9.6}$$

1. We start with the inequality $v \le \hat{v}$. For $t \le u \le T$, we set

$$z := \hat{v}_s(t,s),\ \alpha_u := (\partial_t + \mathcal{L})\hat{v}_s(u,S_u),\ \Gamma_u := S_u\hat{v}_{ss}(u,S_u),$$

and we claim that

$$(\alpha,\Gamma) \in \mathcal{B}_t \text{ and } \Gamma \le \overline{\Gamma}, \tag{9.7}$$

so that the corresponding control $\nu = (y,\alpha,\Gamma) \in \overline{\mathcal{A}}_t(\overline{\Gamma})$. Before verifying this claim, let us complete the proof of the required inequality. Since $g \le \hat{g}$, we have

$$\begin{aligned} g(S_T) \le \hat{g}(S_T) &= \hat{v}(T,S_T) \\ &= \hat{v}(t,S_t) + \int_t^T (\partial_t + \mathcal{L})\hat{v}(u,S_u)\mathrm{d}u + \hat{v}_s(u,S_u)\mathrm{d}S_u \\ &= \hat{v}(t,S_t) + \int_t^T Z_u^\nu \mathrm{d}S_u; \end{aligned}$$

in the last step, we applied Itô's formula to $\hat{v}_s$. Now, observe that the right hand side of the previous inequality is the liquidation value of the portfolio started from the initial capital $\hat{v}(t,S_t)$ and using the portfolio strategy ν. By the definition of the superhedging problem (9.3), we conclude that $v \le \hat{v}$.

It remains to prove (9.7). The upper bound on Γ follows from (9.6). As for the lower bound, it is obtained as a direct consequence of condition (9.4). Using again (9.6) and the smoothness of $\hat{v}$, we see that $0 = \{(\partial_t + \mathcal{L})\hat{v}\}_s = (\partial_t + \mathcal{L})\hat{v}_s + \sigma^2 s\hat{v}_{ss}$, so that $\alpha = -\sigma^2\Gamma$ is also bounded.
2. The proof of the reverse inequality $v \ge \hat{v}$ requires much more effort. The main step is the following (*half*) dynamic programming principle.

Lemma 9.5. *Let $y \in \mathbb{R}$, $\nu \in \overline{\mathcal{A}}_t(\overline{\Gamma})$ be such that $Y_T^{y,\nu} \ge g(S_T)$ P–a.s. Then*

$$Y_\theta^{y,\nu} \ge v(\theta,S_\theta),\ \ \mathbb{P}-a.s.$$

for all stopping times θ valued in $[t,T]$.

The proof of this claim is easy and follows from the same argument than the corresponding one in the standard stochastic target problems of the previous chapter.

We continue by stating two lemmas whose proofs rely heavily on the above dynamic programming principle and will be reported later. We denote as usual by v_* the lower-semicontinuous envelope of v.

Lemma 9.6. *The function v_* is a viscosity supersolution of the equation*

$$-(\partial_t + \mathcal{L})v_* \ge 0 \text{ on } [0,T) \times (0,\infty).$$

Lemma 9.7. *The function $s \longmapsto v_*(t,s) - \overline{\Gamma} s \ln s$ is concave for all $t \in [0,T]$.*

Before proceeding to the proof of these results, let us show how the remaining inequality $v \ge \hat{v}$ follows from it. Given a trading strategy in $\overline{\mathcal{A}}_t(\overline{\Gamma})$, the associated wealth process is a square integrable martingale, and therefore a supermartingale. From this, one easily proves that $v_*(T,s) \geq g(s)$. By Lemma 9.7, $v_*(T,\cdot)$ also satisfies requirement (ii) of Lemma 9.3, and therefore

$$v_*(T,\cdot) \geq \hat{g}.$$

In view of Lemma 9.6, v_* is a viscosity supersolution of the equation $-\mathcal{L}v_* = 0$ and $v_*(T,\cdot) = \hat{g}$. Since $\hat{v}$ is a viscosity solution of the same equation, it follows from the classical comparison theorem that $v_* \geq \hat{v}$.

Hence, in order to complete the proof of Theorem 9.4, it remains to prove Lemmas 9.6 and 9.7.

Proof of Lemmas 9.6 and 9.7. We split the argument in several steps:

1. We first show that the problem can be reduced to the case where the controls (α, Γ) are uniformly bounded. For $\varepsilon \in (0,1]$, set

$$\overline{\mathcal{A}}_t^\varepsilon := \left\{ \nu = (y,\alpha,\Gamma) \in \overline{\mathcal{A}}_t(\overline{\Gamma}) : |\alpha(.)| + |\Gamma(.)| \leq \varepsilon^{-1} \right\},$$

and

$$v^\varepsilon(t,S_t) = \inf\left\{ y \;:\; Y_T^{y,\nu} \geq g(S_T)\ P - \text{a.s. for some } \nu \in \overline{\mathcal{A}}_t^\varepsilon \right\}.$$

Let v_*^ε be the lower-semicontinuous envelope of v^ε. It is clear that v^ε also satisfies the dynamic programming equation of Lemma 9.5.

Since

$$v_*(t,s) = \liminf{}_* v^\varepsilon(t,s) = \liminf_{\varepsilon\to 0,(t',s')\to(t,s)} v_*^\varepsilon(t',s'),$$

we shall prove that

$$-(\partial_t + \mathcal{L})v^\varepsilon \geq 0 \quad \text{in the viscosity sense,} \tag{9.8}$$

and the statement of the lemma follows from the classical stability result of Theorem 6.8.

2. We now derive the implications of the dynamic programming principle of Lemma 9.5 applied to v^ε. Let $\varphi \in C^\infty(\mathbb{R}^2)$ and $(t_0,s_0) \in (0,T)\times(0,\infty)$ satisfy

$$0 = (v_*^\varepsilon - \varphi)(t_0,s_0) = \min_{(0,T)\times(0,\infty)} (v_*^\varepsilon - \varphi);$$

in particular, we have $v_*^\varepsilon \geq \varphi$. Choose a sequence $(t_n,s_n) \to (t_0,s_0)$ so that $v^\varepsilon(t_n,s_n)$ converges to $v_*^\varepsilon(t_0,s_0)$. For each n, by the definition of v^ε and the dynamic programming, there are $y_n \in [v^\varepsilon(t_n,s_n), v^\varepsilon(t_n,s_n) + 1/n]$, hedging strategies $\nu_n = (z_n,\alpha_n,\Gamma_n) \in \overline{\mathcal{A}}_{t_n}^\varepsilon$ satisfying

$$Y_{\theta_n}^{y_n,\nu_n} - v^\varepsilon(\theta_n, S_{\theta_n}) \geq 0$$

for every stopping time θ_n valued in $[t_n, T]$. Since $v^\varepsilon \geq v_*^\varepsilon \geq \varphi$,

$$y_n + \int_{t_n}^{\theta_n} Z_u^{\nu_n} \mathrm{d}S_u - \varphi\left(\theta_n, S_{\theta_n}\right) \geq 0.$$

Observe that

$$\beta_n \;:=\; y_n - \varphi(t_n, s_n) \longrightarrow 0 \;\text{ as } n \longrightarrow \infty.$$

By Itô's formula, this provides

$$M_{\theta_n}^n \leq D_{\theta_n}^n + \beta_n, \tag{9.9}$$

where

$$M_t^n := \int_0^t \left[\varphi_s(t_n + u, S_{t_n+u}) - Y_{t_n+u}^{\nu_n}\right] \mathrm{d}S_{t_n+u}$$

$$D_t^n := -\int_0^t (\partial_t + \mathcal{L})\varphi(t_n + u, S_{t_n+u})\mathrm{d}u\ .$$

We now choose conveniently the stopping time θ_n. For all $h > 0$, define the stopping time

$$\theta_n := (t_n + h) \wedge \inf\{u > t_n \;:\; |\ln(S_u/s_n)| \geq 1\}.$$

3. By the smoothness of $\mathcal{L}\varphi$, the integrand in the definition of M^n is bounded up to the stopping time θ_n and therefore taking expectation in (9.9) provides

$$-E_{t_n, s_n}\left[\int_0^{t\wedge\theta_n} (\partial_t + \mathcal{L})\varphi(t_n + u, S_{t_n+u})\mathrm{d}u\right] \geq -\beta_n$$

We now send n to infinity, divide by h, and take the limit as $h \searrow 0$. The required result follows by dominated convergence.

4. It remains to prove Lemma 9.7. The key-point is the following result, which is a consequence of Theorem 5.5.

Lemma 9.8. *Let* $\left(\{a_u^n,\ u \geq 0\}\right)_n$ *and* $\left(\{b_u^n,\ u \geq 0\}\right)_n$ *be two sequences of real-valued, progressively measurable processes that are uniformly bounded in* n. *Let* (t_n, s_n) *be a sequence in* $[0, T)\times(0, \infty)$ *converging to* $(0, s)$ *for some* $s > 0$. *Suppose that*

$$M_{t\wedge\tau_n}^n := \int_{t_n}^{t_n+t\wedge\tau_n} \left(\zeta_n + \int_{t_n}^u a_r^n \mathrm{d}r + \int_{t_n}^u b_r^n \mathrm{d}S_r\right) \mathrm{d}S_u$$
$$\leq \beta_n + Ct \wedge \tau_n$$

for some real numbers $(\zeta_n)_n$, $(\beta_n)_n$, and stopping times $(\tau_n)_n \geq t_n$. Assume further that, as n tends to infinity,

$$\beta_n \longrightarrow 0 \text{ and } t \wedge \tau_n \longrightarrow t \wedge \tau_0 \quad P - a.s.,$$

where τ_0 is a strictly positive stopping time. Then

(i) $\lim_{n\to\infty} \zeta_n = 0$.
(ii) $\lim_{u\to 0} \text{essinf}_{0\leq r\leq u}\ b_u \leq 0$, *where b be a weak limit process of $(b_n)_n$.*

5. We now complete the proof of Lemma 9.7. We start exactly as in the previous proof by reducing the problem to the case of uniformly bounded controls, and writing the dynamic programming principle on the value function v^ε.

   By a further application of Itô's lemma, we see that

$$M_n(t) = \int_0^t \left(\zeta_n + \int_0^u a_r^n \mathrm{d}r + \int_0^u b_r^n \mathrm{d}S_{t_n+r}\right) \mathrm{d}S_{t_n+u} \tag{9.10}$$

where

$$\begin{aligned}
\zeta_n &:= \varphi_s(t_n, s_n) - z_n,\\
a^n(r) &:= (\partial_t + \mathcal{L})\varphi_s(t_n + r, S_{t_n+r}) - \alpha^n_{t_n+r} l,\\
b_r^n &:= \varphi_{ss}(t_n + r, S_{t_n+r}) - \frac{\Gamma^n_{t_n+r}}{S_{t_n+r}}.
\end{aligned}$$

Observe that the processes $a^n_{.\wedge\theta_n}$ and $b^n_{.\wedge\theta_n}$ are bounded uniformly in n since $(\partial_t + \mathcal{L})\varphi_s$ and φ_{ss} are smooth functions. Also since $(\partial_t + \mathcal{L})\varphi$ is bounded on the stochastic interval $[t_n, \theta_n]$, it follows from (9.9) that

$$M^n_{\theta_n} \leq C\ t \wedge \theta_n\ +\ \beta_n$$

for some positive constant C. We now apply the results of Lemma 9.8 to the martingales M^n. The result is

$$\lim_{n\to\infty} z_n \;=\; \varphi_s(t_0, y_0) \text{ and } \lim_{t\to 0} \text{ess} \inf_{0\leq u\leq t}\ b_t \;\leq\; 0,$$

where b is a weak limit of the sequence (b_n). Recalling that $\Gamma^n(t) \leq \overline{\Gamma}$, this provides that

$$-s_0\varphi_{ss}(t_0, s_0) + \overline{\Gamma} \geq 0.$$

Hence, v^ε_* is a viscosity supersolution of the equation $-s(v_*)_{ss} + \overline{\Gamma} \geq 0$, and the required result follows by the stability result of Theorem 6.8. □

9.1.3 Including the Lower Bound on the Gamma

We now turn to our original problem (9.3) of superhedging under upper and lower bounds on the gamma process.

Following the same intuition as in point 2 of the discussion subsequent to Theorem 9.4, we guess that the value function v should be characterized by the PDE:

$$F(s, \partial_t u, u_{ss}) := \min\left\{-(\partial_t + \mathcal{L})u, \overline{\Gamma} - su_{ss}, su_{ss} - \underline{\Gamma}\right\} = 0,$$

where the first item where the first item of the minimum expression says that the value function should be dominating the Black–Scholes solution, and the two next ones inforce the constraint on the second derivative.

This *first guess* equation is however not elliptic because the third item of the minimum is increasing in u_{ss}. This would divert us from the world of viscosity solutions and the maximum principle. But of course, this is just a guess, and we should expect, as usual in stochastic control, to obtain an elliptic dynamic programming equation.

To understand what is happening in the present situation, we have to go back to the derivation of the DPE from dynamic programming principle in the previous subsection. In particular, we recall that in the proof of Lemmas 9.6 and 9.7, we arrived at the inequality (9.9):

$$M^n_{\theta_n} \leq D^n_{\theta_n} + \beta_n,$$

where

$$D^n_t := -\int_0^t (\partial_t + \mathcal{L})\varphi(t_n + u, S_{t_n+u})\mathrm{d}u,$$

and M_n is given by (9.10), after an additional application of Itô's formula,

$$M_n(t) = \int_0^t \left(\zeta_n + \int_0^u a^n_r \mathrm{d}r + \int_0^u b^n_r \mathrm{d}S_{t_n+r}\right) \mathrm{d}S_{t_n+u}\,,$$

with

$$\begin{aligned}
\zeta_n &:= \varphi_s(t_n, s_n) - z_n,\\
a^n(r) &:= (\partial_t + \mathcal{L})\varphi_s(t_n + r, S_{t_n+r}) - \alpha^n_{t_n+r},\\
b^n_r &:= \varphi_{ss}(t_n + r, S_{t_n+r}) - \frac{\Gamma^n_{t_n+r}}{S_{t_n+r}}.
\end{aligned}$$

To gain more intuition, let us suppress the sequence index n, set $\beta_n = 0$, and take the processes a and b to be constant. Then, we are reduced to the process

$$M(t) = \zeta(S_{t_0+t} - S_{t_0}) + a\int_{t_0}^{t_0+t}(u-t_0)\mathrm{d}S_u + \frac{b}{2}\left((S_{t_0+t} - S_{t_0})^2 - \int_{t_0}^{t}\sigma^2 S_u^2\mathrm{d}u\right).$$

This decomposition reveals many observations:

- The second term should play no role as it is negligible in small time compared to the other ones.
- The requirement $M(.) \leq D(.)$ implies that $b \leq 0$ because otherwise the third term would dominate the other two ones, by the law of iterated logarithm of the Brownian motion, and would converge to $+\infty$ violating the upper bound D. Since $b \leq 0$ and $\underline{\Gamma} \leq \Gamma \leq \overline{\Gamma}$, this provides

$$\underline{\Gamma} \leq s\varphi_{ss} - sb_{t_0} \leq \overline{\Gamma}.$$

- We next observe that, by taking the liminf of the third term, the squared difference $(S_{t_0+t} - S_{t_0})^2$ vanishes. So we may continue as in Step 3 of the proof of Lemmas 9.6 and 9.7, taking expected values, normalizing by h, and sending h to zero. Because of the finite variation component of the third term $\int_{t_0}^{t}\sigma^2 S_u^2 \mathrm{d}u$, this leads to

$$\begin{aligned} 0 &\leq -\partial_t\varphi - \frac{1}{2}\sigma^2 s^2\varphi_{ss} - \frac{b_{t_0}}{2}\sigma^2 s^2 \\ &= -\partial_t\varphi - \frac{1}{2}\sigma^2 s^2(\varphi_{ss} + b_{t_0}). \end{aligned}$$

Collecting the previous inequalities, we arrive at the supersolution property:

$$\hat{F}(s, \partial_t\varphi, \varphi_{ss}) \geq 0,$$

where

$$\hat{F}(s, \partial_t\varphi, \varphi_{ss}) = \sup_{\beta\geq 0} F(s, \partial_t\varphi, \varphi_{ss} + \beta).$$

A remarkable feature of the nonlinearity $\hat{F}$ is that it is elliptic! In fact, it is easy to show that $\hat{F}$ is the smallest elliptic majorant of F. For this reason, we call $\hat{F}$ the elliptic majorant of F.

The above discussion says all about the derivation of the supersolution property. However, more conditions on the set of admissible strategies need to be imposed in order to turn it into a rigorous argument. Once the supersolution property is proved, one also needs to verify that the subsolution property holds true. This also requires to be very careful about the set of admissible strategies. Instead of continuing this example, we shall state without proof the viscosity property, without specifying the precise set of admissible strategies. This question will be studied in details in the subsequent paragraph, where we analzse a general class of second-order stochastic target problems.

Theorem 9.9. *Under a convenient specification of the set $\mathcal{A}(\underline{\Gamma}, \overline{\Gamma})$, the value function v is a viscosity solution of the equation*

$$\hat{F}(s, \partial_t v, v_{ss}) = 0 \text{ on } [0, T) \times \mathbb{R}_+.$$

9.2 Second Order Target Problem

In this section, we introduce the class of second-order stochastic target problems motivated by the hedging problem under gamma constraints of the previous section.

9.2.1 Problem Formulation

The finite time horizon $T \in (0, \infty)$ will be fixed throughout this section. As usual, $\{W_t\}_{t \in [0,T]}$ denotes a d-dimensional Brownian motion on a complete probability space $(\Omega, \mathcal{F}, P)$ and $\mathbb{F} = (\mathcal{F}_t)_{t \in [0,T]}$ the corresponding augmented filtration.

State Processes. We first start from the uncontrolled state process X defined by the stochastic differential equation

$$X_t = x + \int_s^t \mu(X_u) \mathrm{d}u + \int_s^t \sigma(X_u) \mathrm{d}W_u, \quad t \in [s, T].$$

Here, μ and σ are assumed to satisfy the usual Lipschitz and linear growth conditions so as to ensure the existence of a unique solution to the above SDE. We also assume that $\sigma(x)$ is invertible for all $x \in \mathbb{R}^d$.

The control is defined by the $\mathbb{R}^d$-valued process $\{Z_t\}_{t \in [s,T]}$ of the form

$$Z_t = z + \int_s^t A_r \mathrm{d}r + \int_s^t \Gamma_r \mathrm{d}X_r^{s,x}, \quad t \in [s, T], \tag{9.11}$$

$$\Gamma_t = \gamma + \int_s^t a_r \mathrm{d}r + \int_s^t \xi_r \mathrm{d}X_r^{s,x}, \quad t \in [s, T], \tag{9.12}$$

where $\{\Gamma_t\}_{t \in [s,T]}$ takes values in $\mathcal{S}^d$. Notice that both Z and Γ have continuous sample paths, a.s.

Before specifying the exact class of admissible control processes Z, we introduce the controlled state process Y defined by

$$\mathrm{d}Y_t = f(t, X_t^{s,x}, Y_t, Z_t, \Gamma_t)\, \mathrm{d}t + Z_t \circ \mathrm{d}X_t^{s,x}, \quad t \in [s, T), \tag{9.13}$$

with initial data $Y_s = y$. Here, $\circ$ denotes the Fisk–Stratonovich integral. Due to the form of the Z process, this integral can be expressed in terms of standard Itô integral:

$$Z_t \circ \mathrm{d}X_t^{s,x} = Z_t \cdot \mathrm{d}X_t^{s,x} + \frac{1}{2}\mathrm{Tr}[\sigma^{\mathrm{T}}\sigma \Gamma_t]\mathrm{d}t.$$

The function $f : [0,T) \times \mathbb{R}^d \times \mathbb{R} \times \mathbb{R}^d \times \mathcal{S}^d \to \mathbb{R}$, appearing in the controlled state equation (9.13), is assumed to satisfy the following Lipschitz and growth conditions:

(A1) *For all $N > 0$, there exists a constant F_N such that*

$$|f(t,x,y,z,\gamma) - f(t,x,y',z,\gamma)| \leq F_N|y - y'|$$

for all $(t,x,y,z,\gamma) \in [0,T] \times \mathbb{R}^d \times \mathbb{R} \times \mathbb{R}^d \times \mathcal{S}^d$, $y' \in \mathbb{R}$ satisfying

$$\max\{|x|, |y|, |y'|, |z|, |\gamma|\} \leq N.$$

(A2) *There exist constants F and $p \geq 0$ such that*

$$|f(t,x,y,z,\gamma)| \leq F(1 + |x|^p + |y| + |z|^p + |\gamma|^p)$$

for all $(t,x,y,z,\gamma) \in [0,T] \times \mathbb{R}^d \times \mathbb{R} \times \mathbb{R}^d \times \mathcal{S}^d$.

(A3) *There exists a constant $c_0 > 0$ such that*

$$f(t,x,y',z,\gamma) - f(t,x,y,z,\gamma) \geq -c_0(y' - y) \text{ for every } y' \geq y,$$

and $(t,x,z,\gamma) \in [0,T) \times \mathbb{R}^d \times \mathbb{R}^d \times \mathcal{S}^d$.

Admissible Control Processes As outlined in Remark 9.1, the control processes must be chosen so as to exclude the possibility of avoiding the impact of the gamma process by approximation.

We shall fix two constants $B, b \geq 0$ throughout, and we refrain from indexing all the subsequent classes of processes by these constants. For $(s,x) \in [0,T] \times \mathbb{R}^d$, we define the norm of an $\mathbb{F}$–progressively measurable process $\{H_t\}_{t\in[s,T]}$ by,

$$\|H\|_{s,x}^{B,b} := \left\| \sup_{s \leq t \leq T} \frac{|H_t|}{1 + |X_t^{s,x}|^B} \right\|_{\mathbb{L}^b}.$$

For all $m > 0$, we denote by $\mathcal{A}_{m,b}^{s,x}$ be the class of all (control) processes Z of the form (9.11), where the processes A, a, ξ are $\mathbb{F}$-progressively measurable and satisfy:

$$\|Z\|_{s,x}^{B,\infty} \leq m, \qquad \|\Gamma\|_{s,x}^{B,\infty} \leq m, \qquad \|\xi\|_{s,x}^{B,2} \leq m, \tag{9.14}$$

$$\|A\|_{s,x}^{B,b} \leq m, \quad \|a\|_{s,x}^{B,b} \leq m. \tag{9.15}$$

The set of admissible portfolio strategies is defined by

$$\mathcal{A}^{s,x} := \bigcup_{b\in(0,1]} \bigcup_{m\geq 0} \mathcal{A}^{s,x}_{m,b}. \tag{9.16}$$

The Stochastic Target Problem. Let $g : \mathbb{R}^d \to \mathbb{R}$ be a continuous function satisfying the linear growth condition,

(A4) *g is continuous, and there exist constants G and p such that*

$$|g(x)| \leq G(1+|x|^p) \quad \textit{for all } x \in \mathbb{R}^d .$$

For $(s,x) \in [0,T] \times \mathbb{R}^d$, we define:

$$V(s,x) := \inf\left\{y \in \mathbb{R} : Y_T^{s,x,y,Z} \geq g(X_T^{s,x}), \mathbb{P}-\text{a.s. for some } Z \in \mathcal{A}^{s,x}\right\}. \tag{9.17}$$

9.2.2 The Geometric Dynamic Programming

As usual, the key ingredient in order to obtain a PDE satisfied by our value function V is the derivation of a convenient dynamic programming principle obtained by allowing the time origin to move. In the present context, we have the following statement which is similar to the case of standard stochastic target problems.

Theorem 9.10. *For any $(s,x) \in [0,T) \times \mathbb{R}^d$, and a stopping time $\tau \in [s,T]$,*

$$V(s,x) = \inf\left\{y \in \mathbb{R} : Y_\tau^{s,x,y,Z} \geq V\left(\tau, X_\tau^{s,x}\right), \ \mathbb{P}-a.s. \textit{ for some } Z \in \mathcal{A}^{s,x}\right\}. \tag{9.18}$$

The proof of this result can be consulted in [37]. Because the processes Z and Γ are not allowed to jump, the proof is more involved than in the standard stochastic target case, and uses crucially the nature of the above defined class of admissible strategies $\mathcal{A}^{s,x}$.

To derive the dynamic programming equation, we will split the geometric dynamic programming principle in the following two claims:

(GDP1) *For all $\varepsilon > 0$, there exist $y_\varepsilon \in [V(s,x), V(s,x)+\varepsilon]$ and $Z_\varepsilon \in \mathcal{A}^{s,x}$ s.t.*

$$Y_\theta^{s,x,y_\varepsilon,Z_\varepsilon} \geq V\left(\theta, X_\theta^{s,x}\right), \ \mathbb{P}-\text{a.s.} \tag{9.19}$$

(GDP2) *For all $y < V(s,x)$ and every $Z \in \mathcal{A}^{s,x}$,*

$$\mathbb{P}\left[Y_\theta^{s,x,y,Z} \geq V\left(\theta, X_\theta^{s,x}\right)\right] < 1. \tag{9.20}$$

Notice that (9.18) is equivalent to (GDP1) and (GDP2). We shall prove that (GDP1) and (GDP2) imply that the value function V is a viscosity supersolution and subsolution, respectively, of the corresponding dynamic programming equation.

9.2.3 The Dynamic Programming Equation

Similar to the problem of hedging under gamma constraints, the dynamic programming equation corresponding to our second-order target problem is obtained as the parabolic envelope of the first guess equation:

$$-\partial_t v + \hat{f}\left(., v, Dv, D^2 v\right) = 0 \text{ on } [0, T) \times \mathbb{R}^d, \tag{9.21}$$

where

$$\hat{f}(t, x, y, z, \gamma) := \sup_{\beta \in \mathcal{S}_+^d} f(t, x, y, z, \gamma + \beta) \tag{9.22}$$

is the smallest majorant of f which is nonincreasing in the γ argument and is called the parabolic envelope of f. In the following result, we denote by V^* and V_* the upper- and lower-semicontinuous envelopes of V:

$$V_*(t,x) := \liminf_{(t',x') \to (t,x)} V(t', x') \text{ and } V^*(t,x) := \limsup_{(t',x') \to (t,x)} V(t', x')$$

for $(t, x) \in [0, T] \times \mathbb{R}^d$.

Theorem 9.11. *Assume that V is locally bounded, and let conditions (A1–A4) hold true. Then V is a viscosity solution of the dynamic programming equation* (9.21) *on $[0, T] \times \mathbb{R}^d$, i.e., V_* and V^* are, respectively, viscosity super-solution and subsolution of* (9.21).

Proof of the Viscosity Subsolution Property. Let $(t_0, x_0) \in \mathbf{Q}$ and $\varphi \in C^\infty(\mathbf{Q})$ be such that

$$0 = (V^* - \varphi)(t_0, x_0) > (V^* - \varphi)(t, x) \quad \text{for } \mathbf{Q} \ni (t, x) \neq (t_0, x_0). \tag{9.23}$$

In order to show that V^* is a subsolution of (9.21), we assume to the contrary, i.e., suppose that there is $\beta \in \mathcal{S}_+^d$ satisfying

$$-\frac{\partial \varphi}{\partial t}(t_0, x_0) + f\left(t_0, x_0, \varphi(t_0, x_0), D\varphi(t_0, x_0), D^2\varphi(t_0, x_0) + \beta\right) > 0. \tag{9.24}$$

We will then prove the sub-solution property by contradicting (GDP2).

(i) Set

$$\psi(t,x) := \varphi(t,x) + \beta(x-x_0)\cdot(x-x_0),$$
$$h(t,x) := -\frac{\partial\psi}{\partial t}(t,x) + f\left(t,x,\psi(t,x),D\psi(t,x),D^2\psi(t,x)\right).$$

In view of (9.24), $h(t_0,x_0) > 0$. Since the nonlinearity f is continuous and φ is smooth, the subset

$$\mathcal{N} := \{(t,x) \in \mathbf{Q} \cap B_1(t_0,x_0) : \; h(t,x) > 0\}$$

is an open bounded neighborhood of (t_0,x_0). Here, $B_1(t_0,x_0)$ is the unit ball of $\mathbf{Q}$ centered at (t_0,x_0). Since (t_0,x_0) is defined by (9.23) as the point of strict maximum of the difference $(V^* - \varphi)$, we conclude that

$$-\eta := \max_{\partial\mathcal{N}}(V^* - \varphi) < 0. \tag{9.25}$$

Next, we fix $\lambda \in (0,1)$, and choose $\left(\hat{t},\hat{x}\right)$ so that

$$\left(\hat{t},\hat{x}\right) \in \mathcal{N}, \;\; |\hat{x} - x_0| \le \lambda\eta, \text{ and } \left|V\left(\hat{t},\hat{x}\right) - \varphi\left(\hat{t},\hat{x}\right)\right| \;\le\; \lambda\eta. \tag{9.26}$$

Set $\hat{X} := X^{\hat{t},\hat{x}}$ and define a stopping time by

$$\theta := \inf\left\{t \ge \hat{t} : \; (t,\hat{X}_t) \notin \mathcal{N}\right\}.$$

Then, $\theta > \hat{t}$. The path-wise continuity of $\hat{X}$ implies that $(\theta,\hat{X}_\theta) \in \partial\mathcal{N}$. Then, by (9.25),

$$V^*(\theta,\hat{X}_\theta) \le \varphi(\theta,\hat{X}_\theta) - \eta. \tag{9.27}$$

(ii) Consider the control process

$$\hat{z} := D\psi\left(\hat{t},\hat{x}\right), \; \hat{A}_t := \mathcal{L}D\psi(t,\hat{X}_t)\mathbf{1}_{[\hat{t},\theta)}(t) \text{ and } \hat{\Gamma}_t := D^2\psi(t,\hat{X}_t)\mathbf{1}_{[\hat{t},\theta)}(t)$$

so that, for $t \in \left[\hat{t},\theta\right]$,

$$\hat{Z}_t := \hat{z} + \int_{\hat{t}}^t \hat{A}_r \mathrm{d}r + \int_{\hat{t}}^t \hat{\Gamma}_r \mathrm{d}\hat{X}_r \;=\; D\psi(t,\hat{X}_t).$$

Since $\mathcal{N}$ is bounded and φ is smooth, we directly conclude that $\hat{Z} \in \mathcal{A}^{\hat{t},\hat{x}}$.

(iii) Set $\hat{y} < V(\hat{t},\hat{x})$, $\hat{Y}_t := Y_t^{\hat{t},\hat{x},\hat{y},\hat{Z}}$ and $\hat{\Psi}_t := \psi(t,\hat{X}_t)$. Clearly, the process Ψ is bounded on $[\hat{t},\theta]$. For later use, we need to show that the process $\hat{Y}$ is also bounded. By definition, $\hat{Y}_{\hat{t}} < \Psi_{\hat{t}}$. Consider the stopping times

$$\tau_0 := \inf\left\{t \ge \hat{t} : \; \Psi_t = \hat{Y}_t\right\},$$

and, with $N := \eta^{-1}$,

$$\tau_\eta := \inf\left\{t \geq \hat{t} : \hat{Y}_t = \Psi_t - N\right\}.$$

We will show that for a sufficiently large N, both $\tau_0 = \tau_\eta = \theta$. This proves that as Ψ, $\hat{Y}$ is also bounded on $[\hat{t}, \theta]$.

Set $\hat{\theta} := \theta \wedge \tau_0 \wedge \tau_\eta$. Since both processes $\hat{Y}$ and Ψ solve the same stochastic differential equation, it follows from the definition of $\mathcal{N}$ that for $t \in [\hat{t}, \hat{\theta}]$

$$\begin{aligned} \mathrm{d}\left(\Psi_t - \hat{Y}_t\right) &= \left[\frac{\partial \psi}{\partial t}(t, \hat{X}_t) - f\left(t, \hat{X}_t, \hat{Y}_t, \hat{Z}_t, \hat{\Gamma}_t\right)\right] \mathrm{d}t \\ &\leq \left[f\left(t, \hat{X}_t, \Psi_t, \hat{Z}_t, \hat{\Gamma}_t\right) - f\left(t, \hat{X}_t, \hat{Y}_t, \hat{Z}_t, \hat{\Gamma}_t\right)\right] \mathrm{d}t \\ &\leq F_N\left(\Psi_t - \hat{Y}_t\right) \mathrm{d}t\,, \end{aligned}$$

by the local Lipschitz property (A1) of f. Then

$$0 \leq \Psi_{\hat{\theta}} - \hat{Y}_{\hat{\theta}} \leq \left(\Psi_{\hat{t}} - \hat{Y}_{\hat{t}}\right) \mathrm{e}^{F_N T} \leq \frac{1}{2}\|\beta\|\lambda^2 \mathrm{e}^{F_N T}\eta^2, \qquad (9.28)$$

where the last inequality follows from (9.26). This shows that, for λ sufficiently small, $\hat{\theta} < \tau_\eta$, and therefore the difference $\Psi - \hat{Y}$ is bounded. Since Ψ is bounded, this implies that $\hat{Y}$ is also bounded for small η.

(iv) In this step we will show that for any initial data

$$\hat{y} \in [V(\hat{t}, \hat{x}) - \lambda\eta, V(\hat{t}, \hat{x})),$$

we have $\hat{Y}_\theta \geq V(\theta, X_\theta)$. This inequality is in contradiction with (GDP2) as $\hat{Y}_{\hat{t}} = \hat{y} < V(\hat{t}, \hat{x})$. This contradiction proves the subsolution property.

Indeed, using $\hat{y} \geq V(\hat{t}, \hat{x}) - \lambda\eta$ and $V \leq V^* \leq \varphi$ together with (9.25) and (9.26), we obtain the following sequence of inequalities:

$$\begin{aligned} \hat{Y}_\theta - V(\theta, \hat{X}_\theta) &\geq \hat{Y}_\theta - \varphi(\theta, \hat{X}_\theta) + \eta \\ &= [\hat{y} - \varphi(\hat{t}, \hat{x}) + \eta] + \int_{\hat{t}}^{\theta} \left[\mathrm{d}\hat{Y}_t - \mathrm{d}\varphi(t, \hat{X}_t)\right] \\ &\geq \eta(1 - 2\lambda) + \int_{\hat{t}}^{\theta} \left[f\left(t, \hat{X}_t, \hat{Y}_t, \hat{Z}_t, \hat{\Gamma}_t\right) \mathrm{d}t \right. \\ &\qquad\qquad \left. + \hat{Z}_t \circ \mathrm{d}\hat{X}_t - \mathrm{d}\varphi(t, \hat{X}_t)\right] \\ &\geq \eta(1 - 2\lambda) + \frac{1}{2}\beta\left(\hat{X}_\theta - \hat{x}\right) \cdot \left(\hat{X}_\theta - \hat{x}\right) \end{aligned}$$

$$+\int_{\hat{t}}^{\theta}\left[f\left(t,\hat{X}_t,\hat{Y}_t,\hat{Z}_t,\hat{\Gamma}_t\right)\mathrm{d}t+\hat{Z}_t\circ\mathrm{d}\hat{X}_t-\mathrm{d}\psi(t,\hat{X}_t)\right]$$

$$\geq \eta(1-2\lambda)+\int_{\hat{t}}^{\theta}\left[f\left(t,\hat{X}_t,\hat{Y}_t,\hat{Z}_t,\hat{\Gamma}_t\right)\mathrm{d}t\right.$$
$$\left.+\hat{Z}_t\circ\mathrm{d}\hat{X}_t-\mathrm{d}\psi(t,\hat{X}_t)\right],$$

where the last inequality follows from the nonnegativity of the symmetric matrix β. We next use Itô's formula and the definition of $\mathcal{N}$ to arrive at

$$\hat{Y}_\theta-V(\theta,\hat{X}_\theta)\geq\eta(1-2\lambda)+\int_{\hat{t}}^{\theta}\left[f(t,\hat{X}_t,\hat{Y}_t,\hat{Z}_t,\hat{\Gamma}_t)-f(t,\hat{X}_t,\Psi_t,\hat{Z}_t,\hat{\Gamma}_t)\right]\mathrm{d}t.$$

In the previous step, we prove that $\hat{Y}$ and Ψ are bounded, say by N. Since the nonlinearity f is locally bounded, we use the estimate (9.28) to conclude that

$$\hat{Y}_\theta-V\left(\theta,\hat{X}_\theta\right)\geq\eta(1-2\lambda)-\frac{1}{2}\|\beta\|TF_N\mathrm{e}^{F_NT}\lambda^2\eta^2\ \geq 0$$

for all sufficiently small λ. This is in contradiction with (GDP2). Hence, the proof of the viscosity subsolution property is complete.

Proof of the Viscosity Supersolution Property. We first approximate the value function by

$$V^m(s,x):=\inf\{y\in\mathbb{R}\mid\exists Z\in\mathcal{A}_m^{s,x}\text{ so that }Y_T^{s,x,y,Z}\geq g(X_T^{s,x}),\ a.s.\}.$$

Then, similar to (9.20), we can prove the following analogue statement of (GDP1) for V^m:

(GDP1m) *For every* $\varepsilon>0$ *and stopping time* $\theta\in[s,T]$, *there exist* $Z_\varepsilon\in\mathcal{A}_m^{s,x}$ *and* $y_\varepsilon\in[V^m(s,x),V^m(s,x)+\varepsilon]$ *such that* $Y_\theta^{s,x,y_\varepsilon,Z_\varepsilon}\geq V^m\left(\theta,X_\theta^{s,x}\right)$.

Lemma 9.12. V_*^m *is a viscosity supersolution of* (9.21). *Consequently,* V_* *is a viscosity supersolution of* (9.21).

Proof. Choose $(t_0,x_0)\in[s,T)\times\mathbb{R}^d$ and $\varphi\in C^\infty([s,T)\times\mathbb{R}^d)$ such that

$$0=(V_{*,s}^m-\varphi)(t_0,x_0)=\min_{(t,x)\in[s,T)\times\mathbb{R}^d}(V_{*,s}^m-\varphi)(t,x)\,.$$

Let $(t_n,x_n)_{n\geq1}$ be a sequence in $[s,T)\times\mathbb{R}^d$ such that $(t_n,x_n)\to(t_0,x_0)$ and $V^m(t_n,x_n)\to V_{*,s}^m(t_0,x_0)$. There exist positive numbers $\varepsilon_n\to0$ such that for $y_n=V^m(t_n,x_n)+\varepsilon_n$, there exists $Z^n\in\mathcal{A}_m^{t_n,x_n}$ with

$$Y_T^n\geq g(X_T^n),$$

where we use the compact notation $(X^n, Y^n) = (X^{t_n,x_n}, Y^{t_n,x_n,y_n,Z^n})$ and

$$Z^n_r = z_n + \int_{t_n}^r A^n_u \mathrm{d}u + \int_{t_n}^r \Gamma^n_u \mathrm{d}X^n_u,$$

$$\Gamma^n_r = \gamma_n + \int_{t_n}^r a^n_u \mathrm{d}u + \int_{t_n}^r \xi^n_u \mathrm{d}X^n_u, \quad r \in [t_n, T].$$

Moreover, $|z_n|, |\gamma_n| \leq m(1 + |x_n|^p)$ by assumption (9.14). Hence, by passing to a subsequence, we can assume that $z_n \to z_0 \in \mathbb{R}^d$ and $\gamma_n \to \gamma_0 \in \mathcal{S}^d$. Observe that $\alpha_n := y_n - \varphi(t_n, x_n) \to 0$. We choose a decreasing sequence of numbers $\delta_n \in (0, T - t_n)$ such that $\delta_n \to 0$ and $\alpha_n/\delta_n \to 0$. By (GDP1m),

$$Y^n_{t_n+\delta_n} \geq V^m\left(t_n + \delta_n, X^n_{t_n+\delta_n}\right),$$

and therefore

$$Y^n_{t_n+\delta_n} - y_n + \alpha_n \geq \varphi\left(t_n + \delta_n, X^n_{t_n+\delta_n}\right) - \varphi(t_n, x_n),$$

which, after two applications of Itô's formula, becomes

$$\begin{aligned} \alpha_n &+ \int_{t_n}^{t_n+\delta_n} (f(r, X^n_r, Y^n_r, Z^n_r, \Gamma^n_r) - \varphi_t(r, X^n_r))\mathrm{d}r \\ &+ \left(z_n - D\varphi(t_n, x_n)\right) \cdot \left(X^n_{t_n+\delta_n} - x_n\right) \\ &+ \int_{t_n}^{t_n+\delta_n} \left(\int_{t_n}^r [A^n_u - \mathcal{L}D\varphi(u, X^n_u)]\mathrm{d}u\right)' \circ \mathrm{d}X^n_r \\ &+ \int_{t_n}^{t_n+\delta_n} \left(\int_{t_n}^r [\Gamma^n_u - D^2\varphi(u, X^n_u)]\mathrm{d}X^n_u\right)' \circ \mathrm{d}X^n_r \geq 0. \end{aligned} \tag{9.29}$$

It is shown in Lemma 9.13 that the sequence of random vectors

$$\begin{pmatrix} \delta_n^{-1} \int_{t_n}^{t_n+\delta_n} [f(r, X^n_r, Y^n_r, Z^n_r, \Gamma^n_r) - \varphi_t(r, X^n_r)]\mathrm{d}r \\ \delta_n^{-1/2}[X^n_{t_n+\delta_n} - x_n] \\ \delta_n^{-1} \int_{t_n}^{t_n+\delta_n} \left(\int_{t_n}^r [A^n_u - \mathcal{L}D\varphi(u, X^n_u)]\mathrm{d}u\right)' \circ \mathrm{d}X^n_r \\ \delta_n^{-1} \int_{t_n}^{t_n+\delta_n} \left(\int_{t_n}^r [\Gamma^n_u - D^2\varphi(u, X^n_u)]\mathrm{d}X^n_u\right)' \circ \mathrm{d}X^n_r \end{pmatrix}, \ n \geq 1, \tag{9.30}$$

converges in distribution to

$$\begin{pmatrix} f(t_0, x_0, \varphi(t_0, x_0), z_0, \gamma_0) - \varphi_t(t_0, x_0) \\ \sigma(x_0)W_1 \\ 0 \\ \frac{1}{2}W_1 \cdot \sigma(x_0)^{\mathrm{T}}\left(\gamma_0 - D^2\varphi(t_0, x_0)\right)\sigma(x_0)W_1 \end{pmatrix}. \tag{9.31}$$

Set $\eta_n = |z_n - D\varphi(t_n, x_n)|$, and assume $\delta_n^{-1/2}\eta_n \to \infty$ along a subsequence. Then, along a further subsequence, $\eta_n^{-1}(z_n - D\varphi(t_n, x_n))$ converges to some $\eta_0 \in \mathbb{R}^d$ with

$$|\eta_0| = 1. \tag{9.32}$$

Multiplying inequality (9.29) with $\delta_n^{-1/2}\eta_n^{-1}$ and passing to the limit yields

$$\eta_0 \cdot \sigma(x_0)W_1 \geq 0,$$

which, since $\sigma(x_0)$ is invertible, contradicts (9.32). Hence, the sequence $(\delta_n^{-1/2}\eta_n)$ has to be bounded, and therefore, possibly after passing to a subsequence,

$$\delta_n^{-1/2}[z_n - D\varphi(t_n, x_n)] \quad \text{converges to some } \xi_0 \in \mathbb{R}^d.$$

It follows that $z_0 = D\varphi(t_0, x_0)$. Moreover, we can divide inequality (9.29) by δ_n and pass to the limit to get

$$\begin{aligned} & f(t_0, x_0, \varphi(t_0, x_0), D\varphi(t_0, x_0), \gamma_0) - \varphi_t(t_0, x_0) \\ & \quad + \xi_0 \cdot \sigma(x_0)W_1 + \frac{1}{2}W_1 \cdot \sigma(x_0)^{\mathrm{T}}[\gamma_0 - D^2\varphi(t_0, x_0)]\sigma(x_0)W_1 \geq 0. \end{aligned} \tag{9.33}$$

Since the support of the random vector W_1 is $\mathbb{R}^d$, it follows from (9.33) that

$$\begin{aligned} & f(t_0, x_0, \varphi(t_0, x_0), D\varphi(t_0, x_0), \gamma_0) - \varphi_t(t_0, x_0) \\ & \quad + \xi_0 \cdot \sigma(x_0)w + \frac{1}{2}w \cdot \sigma(x_0)^{\mathrm{T}}[\gamma_0 - D^2\varphi(t_0, x_0)]\sigma(x_0)w \geq 0, \end{aligned}$$

for all $w \in \mathbb{R}^d$. This shows that

$$f(t_0, x_0, \varphi(t_0, x_0), D\varphi(t_0, x_0), \gamma_0) - \varphi_t(t_0, x_0) \geq 0 \text{ and } \beta := \gamma_0 - D^2\varphi(t_0, x_0) \geq 0,$$

and therefore,

$$-\varphi_t(t_0, x_0) + \sup_{\beta \in \mathcal{S}_+^d} f(t_0, x_0, \varphi(t_0, x_0), D\varphi(t_0, x_0), D^2\varphi(t_0, x_0) + \beta) \geq 0\,.$$

This proves that V^m is a viscosity supersolution.

Since by definition,

$$V = \inf_m V^m,$$

by the classical stability property of viscosity solutions, V_* is also a viscosity supersolution of the DPE (9.21). In fact, this passage to the limit does not fall exactly into the stability result of Theorem 6.8, but its justification follows the lines of the proof of stability, the interested reader can find the detailed argument in Corollary 5.5 in [12]. □

Lemma 9.13. *The sequence of random vectors* (9.30), *on a subsequence, converges in distribution to* (9.31).

Proof. Define a stopping time by

$$\tau_n := \inf\{r \geq t_n : X_r^n \notin B_1(x_0)\} \wedge (t_n + \delta_n),$$

where $B_1(x_0)$ denotes the open unit ball in $\mathbb{R}^d$ around x_0. It follows from the fact that $x_n \to x_0$ that

$$P[\tau_n < t_n + \delta_n] \to 0$$

so that in (9.30), we may replace the upper limits of the integrations by τ_n instead of $t_n + \delta_n$.

Therefore, in the interval $[t_n, \tau_n]$, the process X^n is bounded. Moreover, in view of (9.15) so are Z^n, Γ^n, and ξ^n.

Step 1. The convergence of the second component of (9.30) is straightforward and the details are exactly as in Lemma 4.4 [13].

Step 2. Let B be as in (9.14). To analyze the other components, set

$$A^{n,*} := \sup_{u \in [t_n, T]} \frac{|A_u^n|}{1 + |X_u^n|^B},$$

so that, by (9.15),

$$\|A^{n,*}\|_{L^{(1/m)}(\Omega, \mathbb{P})} \leq m. \tag{9.34}$$

Moreover, since on the interval $[t_n, \tau_n]$, X^n is uniformly bounded by a deterministic constant $C(x_0)$ depending only on x_0,

$$\left|A_u^n\right| \leq C(x_0)\, A^{n,*} \leq C(x_0) m, \qquad \forall\, u \in [t_n, \tau_n].$$

(Here and below, the constant $C(x_0)$ may change from line to line.) We define $a^{n,*}$ similarly. Then, it also satisfies the above bounds as well. In view of (9.15), also $a^{n,*}$ satisfies (9.34). Moreover, using (9.14), we conclude that ξ_u^n is uniformly bounded by m.

Step 3. Recall that $\mathrm{d}\Gamma_u^n = a_u^n \mathrm{d}u + \xi_u^n \mathrm{d}X_u^n$, $\Gamma_{t_n}^n = \gamma_n$. Using the notation and the estimates of the previous step, we directly calculate that

$$\begin{aligned}
\sup_{t \in [t_n, \tau_n]} |\Gamma_t^n - \gamma_n| &\leq C(x_0)\delta_n a^{n,*} + \left| \int_{t_n}^{\tau_n} \xi_u^n \cdot \mu^n \mathrm{d}u \right| + \left| \int_{t_n}^{\tau_n} \xi_u^n \sigma(X_u^n) \mathrm{d}W_u \right| \\
&:= I_1^n + I_2^n + I_3^n.
\end{aligned}$$

Then,

$$E[(I_3^n)^2] \leq E\left(\int_{t_n}^{\tau_n} |\xi_u^n|^2 |\sigma|^2 \mathrm{d}u \right) \leq \delta_n\, m^2 C(x_0)^2.$$

Hence, I_3^n converges to zero in L^2. Therefore, it also converges almost surely on a subsequence. We prove the convergence of I_2^n using similar estimates. Since $a^{n,*}$ satisfies (9.34),

$$E[(I_1^n)^{(1/m)}] \leq (C(x_0)\delta_n)^{(1/m)}\ E[|a^{n,*}|^{(1/m)}] \leq (C(x_0)\delta_n)^{(1/m)}\ m.$$

Therefore, I_1^n converges to zero in $L^{(1/m)}$ and consequently on almost surely on a subsequence.

Hence, on a subsequence, Γ_t^n is uniformly continuous. This together with standard techniques used in Lemma 4.4 of [13] proves the convergence of the first component of (9.30).

Step 4. By integration by parts,

$$\int_{t_n}^{\tau_n}\int_{t_n}^{t} A_u^n \mathrm{d}u \mathrm{d}X_t^n = (X_{\tau_n}^n - X_{t_n}^n)\int_{t_n}^{\tau_n} A_u^n \mathrm{d}u - \int_{t_n}^{\tau_n}(X_u^n - X_{t_n}^n)A_u^n \mathrm{d}u.$$

Therefore,

$$\left|\frac{1}{\delta_n}\int_{t_n}^{\tau_n}\int_{t_n}^{t} A_u^n \mathrm{d}u \mathrm{d}X_t^n\right| \leq C(x_0)\sup_{t\in[t_n,\tau_n]}|X_t^n - X_{t_n}^n|\ A^{n,*}.$$

Also X^n is uniformly continuous and $A^{n,*}$ satisfies (9.34). Hence, we can show that the above terms, on a subsequence, almost surely converge to zero. This implies the convergence of the third term.

Step 5. To prove the convergence of the final term, it suffices to show that

$$J^n := \frac{1}{\delta_n}\int_{t_n}^{\tau_n}\int_{t_n}^{t}[\Gamma_u^n - \gamma_n]\mathrm{d}X_u^n \circ \mathrm{d}X_t^n$$

converges to zero. Indeed, since $\gamma_n \to \gamma_0$, this convergence together with the standard arguments of Lemma 4.4 of [13] yields the convergence of the fourth component.

Since on $[t_n, \tau_n]$ X^n is bounded, on this interval $|\sigma(X_t^n)| \leq C(x)$. Using this bound, we calculate that

$$\begin{aligned} E[(J^n)^2] &\leq \frac{C(x_0)^4}{\delta_n^2}\int_{t_n}^{t_n+\delta_n}\int_{t_n}^{t} E\left[\mathbf{1}_{[t_n,\tau_n]}\,|\Gamma_u^n - \gamma_n|^2\right]\mathrm{d}u\,\mathrm{d}t \\ &\leq C(x_0)^4 E\left[\sup_{t\in[t_n,\tau_n]}|\Gamma_u^n - \gamma_n|^2\right] =: C(x_0)^4 E\left[(\mathrm{e}^n)^2\right]. \end{aligned}$$

In Step 3, we proved the almost sure convergence of e^n to zero. Moreover, by (9.14), $|\mathrm{e}^n| \leq m$. Therefore, by dominated convergence, we conclude that J^n converges to zero in L^2. Thus almost everywhere on a subsequence. □

9.3 Superhedging Under Illiquidity Cost

In this section, we analyze the superhedging problem under a more realistic model accounting for the market illiquidity. We refer to [9] for all technical details.

Following Cetin et al. [8] (CJP, hereafter), we account for the liquidity cost by modeling the price process of this asset as a function of the exchanged volume. We thus introduce a supply curve

$$\mathbf{S}\left(S_t, \nu\right),$$

where $\nu \in \mathbb{R}$ indicates the volume of the transaction, the process $S_t = \mathbf{S}\left(S_t, 0\right)$ is the *marginal price process* defined by some given initial condition $S(0)$ together with the Black–Scholes dynamics:

$$\frac{\mathrm{d}S_t}{S_t} = \sigma \mathrm{d}W_t, \tag{9.35}$$

where as usual the prices are discounted, i.e. expressed in the numéraire defined by the nonrisky asset, and the drift is omitted by a change of measure.

The function $\mathbf{S} : \mathbb{R}_+ \times \mathbb{R} \longrightarrow \mathbb{R}$ is assumed to be smooth and increasing in ν. $\mathbf{S}(s, \nu)$ represents the price per share for trading of size ν and marginal price s.

A trading strategy is defined by a pair (Z^0, Z) where Z_t^0 is the position in cash and Z_t is the number of shares held at each time t in the portfolio. As in the previous paragraph, we will take the process Z in the set of admissible strategies $\mathcal{A}^{t,s}$ defined in (9.16), whenever the problem is started at the time origin t with the initial spot price s for the underlying asset.

To motivate the continuous-time model, we start from discrete-time trading strategies. Let $0 = t_0 < \cdots < t_n = T$ be a partition of the time interval $[0, T]$, and denote $\delta\psi(t_i) := \psi(t_i) - \psi(t_{i-1})$ for any function ψ. By the self-financing condition, it follows that

$$\delta Z_{t_i}^0 + \delta Z_{t_i} \mathbf{S}\left(S_{t_i}, \delta Z_{t_i}\right) = 0, \ 1 \le i \le n.$$

Summing up these equalities, it follows from direct manipulations that

$$\begin{aligned}
Z_T^0 + Z_T S_T &= Z_0^0 + Z_0 S_0 - \sum_{i=1}^n \left[\delta Z_{t_i} \mathbf{S}\left(S_{t_i}, \delta Z_{t_i}\right) + \left(Z_0 S_0 - Z_T S_T\right)\right] \\
&= Z_0^0 + Z_0 S_0 - \sum_{i=1}^n \left[\delta Z_{t_i} S_{t_i} + \left(Z_0 S_0 - Z_T S_T\right)\right] \\
&\quad - \sum_{i=1}^n \delta Z_{t_i} \left[\mathbf{S}\left(S_{t_i}, \delta Z_{t_i}\right) - S_{t_i}\right] \\
&= Z_0^0 + Z_0 S_0 + \sum_{i=1}^n Z_{t_{i-1}} \delta S_{t_i} - \sum_{i=1}^n \delta Z_{t_i} \left[\mathbf{S}\left(S_{t_i}, \delta Z_{t_i}\right) - S_{t_i}\right].
\end{aligned} \tag{9.36}$$

Then, the continuous-time dynamics of the process

$$Y := Z^0 + ZS$$

are obtained by taking limits in (9.36) as the time step of the partition shrinks to zero. The last sum term in (9.36) is the term due to the liquidity cost.

Since the function $\nu \longmapsto \mathbf{S}(s,\nu)$ is assumed to be smooth, it follows from the form of the continuous-time process Z in (9.11) that:

$$Y_t = Y_0 + \int_0^t Z_u \mathrm{d}S_u - \int_0^t \frac{4}{S_u\phi(S_u)} \mathrm{d}\langle Z\rangle_u \tag{9.37}$$

$$= Y_0 + \int_0^t Z_u \mathrm{d}S_u - \int_0^t \frac{4}{\ell(S_u)} \Gamma_u^2 \sigma^2(u, S_u) S_u \mathrm{d}u, \tag{9.38}$$

where ℓ is the liquidity function defined by

$$\ell(s) := s\left(\frac{\mathbf{S}}{\partial v}(s,0)\right)^{-1}. \tag{9.39}$$

The above liquidation value of the portfolio exhibits a penalization by a linear term in Γ^2, with coefficient determined by the slope of the order book at the origin. This type of dynamics falls into the general problems analyzed in the previous section.

Remark 9.14. The supply function $\mathbf{S}(s,\nu)$ can be inferred from the data on order book prices. We refer to [10] for a parametric estimation of this model on real financial data.

In the context of the CJP model, we ignore the illiquidity cost at the maturity date T, and we formulate the super-hedging problem by:

$$V(t,s) := \inf\left\{y : Y_T^{y,Z} \geq g(S_T^{t,s}),\ \mathbb{P}-\text{a.s. for some } Z \in \mathcal{A}^{t,s}\right\}. \tag{9.40}$$

Then, the viscosity property for the value function V follows from the results of the previous section. The next result says more as it provides uniqueness.

Theorem 9.15. *Assume that V is locally bounded. Then, the super-hedging cost V is the unique viscosity solution of the PDE problem*

$$-\partial_t V - \frac{1}{2}\sigma^2 s H\big((-\ell) \vee (sV_{ss})\big) = 0,\ \ V(T,.) = g \tag{9.41}$$

$$-C \leq V(t,s) \leq C(1+s),\ (t,s) \in [0,T]\times\mathbb{R}_+,\quad \textit{for some } C > 0, \tag{9.42}$$

where $H(\gamma) := \gamma + \frac{1}{2\ell}\gamma^2$.

We refer to [9] for the proof of uniqueness. We conclude this section by some comments.

Remark 9.16. 1. The PDE (9.41) is very similar to the PDE obtained in the problem of hedging under Gamma constraints. We observe here that $-\ell$ plays the same role as the lower bound $\underline{\Gamma}$ on the Gamma of the portfolio. Therefore, the CJP model induces an endogenous (state-dependent) lower bound on the Gamma of the portfolio defined by ℓ.

2. However, there is no counterpart in (9.41) to the upper bound $\overline{\Gamma}$ which induced the face-lifting of the payoff in the problem of hedging under Gamma constraints.

Chapter 10
Backward SDEs and Stochastic Control

In this chapter, we introduce the notion of backward stochastic differential equation (BSDE hereafter) which allows to relate standard stochastic control to stochastic target problems. More importantly, the general theory in this chapter will be developed in the non-Markovian framework. The Markovian framework of the previous chapters and the corresponding PDEs will be obtained under a specific construction. From this viewpoint, BSDEs can be viewed as the counterpart of PDEs in the non-Markovian framework.

However, by their very nature, BSDEs can only cover the subclass of standard stochastic control problems with uncontrolled diffusion, with corresponding semi-linear DPE. Therefore, a further extension is needed in order to cover the more general class of fully nonlinear PDEs, as those obtained as the DPE of standard stochastic control problems. This can be achieved by means of the notion of second-order BSDEs which are very connected to second-order target problems. We refer to Cheridito et al. [13] and Soner et al. [38] for this extension.

10.1 Motivation and Examples

The first appearance of BSDEs was in the early work of Bismut [6] who was extending the Pontryagin maximum principle of optimality to the stochastic framework. Similar to the deterministic context, this approach introduces the so-called adjoint process defined by a stochastic differential equation combined with a final condition. In the deterministic framework, the existence of a solution to the adjoint equation follows from the usual theory by obvious time inversion. The main difficulty in the stochastic framework is that the adjoint process is required to be adapted to the given filtration, so that one cannot simply solve the existence problem by running the time clock backward.

A systematic study of BSDEs was started by Pardoux and Peng [32]. The motivation was also from optimal control which was an important field of interest

N. Touzi, *Optimal Stochastic Control, Stochastic Target Problems, and Backward SDE*, Fields Institute Monographs 29, DOI 10.1007/978-1-4614-4286-8_10,

for Shige Peng. However, the natural connections with problems in financial mathematics were very quickly realized, see Elkaroui et al. [18]. Therefore, a large development of the theory was achieved in connection with financial applications and crucially driven by the intuition from finance.

10.1.1 The Stochastic Pontryagin Maximum Principle

Our objective in this section is to see how the notion of BSDE appears naturally in the context of the Pontryagin maximum principle. Therefore, we are not intending to develop any general theory about this important question, and we will not make any effort in weakening the conditions for the main statement. We will instead considerably simplify the mathematical framework in order for the main ideas to be as transparent as possible.

Consider the stochastic control problem

$$V_0 := \sup_{\nu \in \mathcal{U}_0} J_0(\nu) \text{ where } J_0(\nu) := \mathbb{E}\left[g(X_T^\nu)\right],$$

the set of control processes $\mathcal{U}_0$ is defined as in Sect. 3.1, and the controlled state process is defined by some initial date X_0 and the SDE with random coefficients:

$$dX_t^\nu = b(t, X_t^\nu, \nu_t)dt + \sigma(t, X_t^\nu, \nu_t)dW_t.$$

Observe that we are not emphasizing the time origin and the position of the state variable X at the time origin. This is a major difference between the dynamic programming approach, developed by the American school, and the Pontryagin maximum principle approach of the Russian school.

For every $u \in U$, we define:

$$L^u(t, x, y, z) := b(t, x, u) \cdot y + \mathrm{Tr}\left[\sigma(t, x, u)^{\mathrm{T}} z\right],$$

so that

$$b(t, x, u) = \frac{\partial L^u(t, x, y, z)}{\partial y} \text{ and } \sigma(t, x, u) = \frac{\partial L^u(t, x, y, z)}{\partial z}.$$

We also introduce the function

$$\ell(t, x, y, z) := \sup_{u \in U} L^u(t, x, y, z),$$

and we will denote by $\mathbb{H}^2$ the space of all $\mathbb{F}$−progressively measurable processes with finite $\mathbb{L}^2\left([0, T] \times \Omega, dt \otimes d\mathbb{P}\right)$−norm.

Theorem 10.1. *Let $\hat{\nu} \in \mathcal{U}_0$ be such that*

(i) *There is a solution $(\hat{Y}, \hat{Z})$ in $\mathbb{H}^2$ of the backward stochastic differential equation:*

$$d\hat{Y}_t = -\nabla_x L^{\hat{\nu}_t}(t, \hat{X}_t, \hat{Y}_t, \hat{Z}_t)dt + Z_t dW_t, \text{ and } \hat{Y}_T = \nabla g(\hat{X}_T) \quad (10.1)$$

where $\hat{X} := X^{\hat{\nu}}$

(ii) *$\hat{\nu}$ satisfies the maximum principle:*

$$L^{\hat{\nu}_t}(t, \hat{X}_t, \hat{Y}_t, \hat{Z}_t) = \ell(t, \hat{X}_t, \hat{Y}_t, \hat{Z}_t). \quad (10.2)$$

(iii) *The functions g and $\ell(t, ., y, z)$ are concave, for fixed t, y, z, and*

$$\nabla_x L^{\hat{\nu}_t}(t, \hat{X}_t, \hat{Y}_t, \hat{Z}_t) = \nabla_x \ell(t, \hat{X}_t, \hat{Y}_t, \hat{Z}_t). \quad (10.3)$$

Then $V_0 = J_0(\hat{\nu})$, i.e., $\hat{\nu}$ is an optimal control for the problem V_0.

Proof. For an arbitrary $\nu \in \mathcal{U}_0$, we compute that

$$\begin{aligned} J_0(\hat{\nu}) - J_0(\nu) &= \mathbb{E}\Big[g(\hat{X}_T) - g(X_T^\nu)\Big] \\ &\geq \mathbb{E}\Big[(\hat{X}_T - X_T^\nu) \cdot \nabla g(\hat{X}_T)\Big] \\ &= \mathbb{E}\Big[(\hat{X}_T - X_T^\nu) \cdot \hat{Y}_T\Big] \end{aligned}$$

by the concavity assumption on g. Using the dynamics of $\hat{X}$ and $\hat{Y}$, this provides

$$\begin{aligned} J_0(\hat{\nu}) - J_0(\nu) &\geq \mathbb{E}\Big[\int_0^T d\{(\hat{X}_T - X_T^\nu) \cdot \hat{Y}_T\}\Big] \\ &= \mathbb{E}\Big[\int_0^T \big(b(t, \hat{X}_t, \hat{\nu}_t) - b(t, X_t^\nu, \nu_t)\big) \cdot \hat{Y}_t dt \\ &\quad -(\hat{X}_t - X_t^\nu) \cdot \nabla_x L^{\hat{\nu}_t}(t, \hat{X}_t, \hat{Y}_t, \hat{Z}_t)dt \\ &\quad +\mathrm{Tr}\Big[\big(\sigma(t, \hat{X}_t, \hat{\nu}_t) - \sigma(t, X_t^\nu, \nu_t)\big)^{\mathrm{T}} \hat{Z}_t\Big] dt\Big] \\ &= \mathbb{E}\Big[\int_0^T \big(L^{\hat{\nu}_t}(t, \hat{X}_t, \hat{Y}_t, \hat{Z}_t) - L^{\nu_t}(t, X_t, \hat{Y}_t, \hat{Z}_t) \\ &\quad -(\hat{X}_t - X_t^\nu) \cdot \nabla_x L^{\hat{\nu}_t}(t, \hat{X}_t, \hat{Y}_t, \hat{Z}_t)\big)dt\Big], \end{aligned}$$

where the diffusion terms have zero expectations because the processes $\hat{Y}$ and $\hat{Z}$ are in $\mathbb{H}^2$. By Conditions (ii) and (iii), this implies that

$$J_0(\hat{\nu}) - J_0(\nu) \geq \mathbb{E}\left[\int_0^T \left(\ell(t, \hat{X}_t, \hat{Y}_t, \hat{Z}_t) - \ell(t, X_t, \hat{Y}_t, \hat{Z}_t)\right.\right.$$
$$\left.\left. -(\hat{X}_t - X_t^\nu) \cdot \nabla_x \ell(t, \hat{X}_t, \hat{Y}_t, \hat{Z}_t)\right)dt\right]$$
$$\geq 0$$

by the concavity assumption on ℓ. □

Let us comment on the conditions of the previous theorem.

- Condition (ii) provides a feedback definition to $\hat{\nu}$. In particular, $\hat{\nu}_t$ is a function of $(t, \hat{X}_t, \hat{Y}_t, \hat{Z}_t)$. As a consequence, the forward SDE defining $\hat{X}$ depends on the backward component $(\hat{Y}, \hat{Z})$. This is a situation of forward-backward stochastic differential equation which will not be discussed in these notes.
- Condition (10.3) in (iii) is satisfied under natural smoothness conditions. In the economic literature, this is known as the envelope theorem.
- Condition (i) states the existence of a solution to the BSDE (10.1), which will be the main focus of the subsequent section.

10.1.2 BSDEs and Stochastic Target Problems

Let us go back to a subclass of the stochastic target problems studied in Chap. 8 defined by taking the state process X independent of the control Z which is assumed to take values in $\mathbb{R}^d$. For simplicity, let $X = W$. Then the stochastic target problem is defined by

$$V_0 := \inf\left\{Y_0 : \ Y_T^Z \geq g(W_T), \ \mathbb{P} - \text{a.s. for some } Z \in \mathbb{H}^2\right\},$$

where the controlled process Y satisfies the dynamics:

$$dY_t^Z = b(t, W_t, Y_t, Z_t)dt + Z_t \cdot dW_t. \tag{10.4}$$

If existence holds for the latter problem, then there would exist a pair (Y, Z) in $\mathbb{H}^2$ such that

$$Y_0 + \int_0^T \left[b(t, W_t, Y_t, Z_t)dt + Z_t \cdot dW_t\right] \geq g(W_T), \ \ \mathbb{P} - \text{a.s.}$$

If in addition equality holds in the latter inequality, then (Y, Z) is a solution of the BSDE defined by (10.4) and the terminal condition $Y_T = g(W_T)$, $\mathbb{P}$–a.s.

10.1.3 BSDEs and Finance

In the Black–Scholes model, we know that any derivative security can be perfectly hedged. The corresponding superhedging problem reduces to a hedging problem, and an optimal hedging portfolio exists and is determined by the martingale representation theorem.

In fact, this goes beyond the Markovian framework to which the stochastic target problems are restricted. To see this, consider a financial market with interest rate process $\{r_t, t \geq 0\}$ and d risky assets with price process defined by

$$\mathrm{d}S_t = S_t \star (\mu_t \mathrm{d}t + \sigma_t \mathrm{d}W_t).$$

Then, under the self-financing condition, the liquidation value of the portfolio is defined by

$$\mathrm{d}Y_t^{\pi} = r_t Y_t^{\pi} \mathrm{d}t + \pi_t \sigma_t (\mathrm{d}W_t + \lambda_t \mathrm{d}t), \tag{10.5}$$

where the risk premium process $\lambda_t := \sigma_t^{-1}(\mu_t - r_t \mathbf{1})$ is assumed to be well defined, and the control process π_t denotes the vector of holdings amounts in the d risky assets at each point in time.

Now let G be a random variable indicating the random payoff of a contract. G is called a *contingent claim*. The hedging problem of G consists in searching for a portfolio strategy $\hat{\pi}$ such that

$$Y_T^{\hat{\pi}} = G, \ \mathbb{P}-\text{a.s.} \tag{10.6}$$

We are then reduced to a problem of solving the BSDE (10.5)–(10.6). This problem can be solved very easily if the process λ is so that the process $\{W_t + \int_0^t \lambda_s \mathrm{d}s, t \geq 0\}$ is a Brownian motion under the so-called equivalent probability measure $\mathbb{Q}$. Under this condition, it suffices to get rid of the linear term in (10.5) by discounting, then $\hat{\pi}$ is obtained by the martingale representation theorem in the present Brownian filtration under the equivalent measure $\mathbb{Q}$.

We finally provide an example where the dependence of Y in the control variable Z is nonlinear. The easiest example is to consider a financial market with different lending and borrowing rates $\underline{r}_t \leq \overline{r}_t$. Then the dynamics of liquidation value of the portfolio (10.5) is replaced by the following SDE:

$$\mathrm{d}Y_t = \pi_t \cdot \sigma_t (\mathrm{d}W_t + \lambda_t \mathrm{d}t)(Y_t - \pi_t \cdot \mathbf{1})^+ \underline{r}_t - (Y_t - \pi_t \cdot \mathbf{1})^- \overline{r}_t. \tag{10.7}$$

As a consequence of the general result of the subsequent section, we will obtain the existence of a hedging process $\hat{\pi}$ such that the corresponding liquidation value satisfies (10.7) together with the hedging requirement (10.6).

10.2 Wellposedness of BSDEs

Throughout this section, we consider a $d-$dimensional Brownian motion W on a complete probability space $(\Omega, \mathcal{F}, \mathbb{P})$, and we denote by $\mathbb{F} = \mathbb{F}^W$ the corresponding augmented filtration.

Given two integers $n, d \in \mathbb{N}$, we consider the mapping

$$f : [0,T] \times \Omega \times \mathbb{R}^n \times \mathbb{R}^{n\times d} \longrightarrow \mathbb{R}$$

that we assume to be $\mathcal{P} \otimes \mathcal{B}(\mathbb{R}^{n+nd})-$measurable, where $\mathcal{P}$ denotes the $\sigma-$algebra generated by predictable processes. In other words, for every fixed $(y,z) \in \mathbb{R}^n \times \mathbb{R}^{n\times d}$, the process $\{f_t(y,z), t \in [0,T]\}$ is $\mathbb{F}-$predictable.

Our interest is on the BSDE:

$$dY_t = -f_t(Y_t, Z_t)dt + Z_t dW_t \text{ and } Y_T = \xi, \ \mathbb{P}-\text{a.s.} \tag{10.8}$$

where ξ is some given $\mathcal{F}_T-$measurable r.v. with values in $\mathbb{R}^n$.

We will refer to (10.8) as BSDE(f, ξ). The map f is called the *generator*. We may also re-write the BSDE (10.8) in the integrated form:

$$Y_t = \xi + \int_t^T f_s(Y_s, Z_s)ds - \int_t^T Z_s dW_s, \ t \le T \ , \ \mathbb{P}-\text{a.s.} \tag{10.9}$$

10.2.1 Martingale Representation for Zero Generator

When the generator $f \equiv 0$, the BSDE problem reduces to the martingale representation theorem in the present Brownian filtration. More precisely, for every $\xi \in \mathbb{L}^2(\mathbb{R}^n, \mathcal{F}_T)$, there is a unique pair process (Y, Z) in $\mathbb{H}^2(\mathbb{R}^n \times \mathbb{R}^{n\times d})$ satisfying (10.8):

$$\begin{aligned} Y_t &:= \mathbb{E}[\xi|\mathcal{F}_t] = \mathbb{E}[\xi] + \int_0^t Z_s dW_s \\ &= \xi - \int_t^T Z_s dW_s. \end{aligned}$$

Here, for a subset E of $\mathbb{R}^k$, $k \in \mathbb{N}$, we denoted by $\mathbb{H}^2(E)$ the collection of all $\mathbb{F}-$progressively measurable $\mathbb{L}^2([0,T] \times \Omega, \text{Leb} \otimes \mathbb{P})-$processes with values in E. We shall frequently simply write $\mathbb{H}^2$ keeping the reference to E implicit.

Let us notice that Y is a uniformly integrable martingale. Moreover, by the Doob's maximal inequality, we have

$$\|Y\|_{\mathcal{S}^2}^2 := \mathbb{E}\left[\sup_{t\le T}|Y_t|^2\right] \le 4\mathbb{E}\left[|Y_T|^2\right] = 4\|Z\|_{\mathbb{H}^2}^2. \tag{10.10}$$

Hence, the process Y is in the space of continuous processes with finite $\mathcal{S}^2$–norm.

10.2.2 BSDEs with Affine Generator

We next consider a scalar BSDE ($n = 1$) with generator

$$f_t(y,z) := a_t + b_t y + c_t \cdot z, \tag{10.11}$$

where $a, b, andc$ are $\mathbb{F}$–progressively measurable processes with values in $\mathbb{R}$, $\mathbb{R}$, and $\mathbb{R}^d$, respectively. We also assume that b, c are bounded and $\mathbb{E}\left[\int_0^T |a_t|^2 \mathrm{d}t\right] < \infty$.

This case is easily handled by reducing to the zero generator case. However, it will play a crucial role for the understanding of BSDEs with generator quadratic in z, which will be the focus of the next chapter.

First, by introducing the equivalent probability $\mathbb{Q} \sim \mathbb{P}$ defined by the density

$$\frac{\mathrm{d}\mathbb{Q}}{\mathrm{d}\mathbb{P}} = \exp\left(\int_0^T c_t \cdot \mathrm{d}W_t - \frac{1}{2}\int_0^T |c_t|^2 \mathrm{d}t\right),$$

it follows from the Girsanov theorem that the process $B_t := W_t - \int_0^t c_s \mathrm{d}s$ defines a Brownian motion under $\mathbb{Q}$. By formulating the BSDE under $\mathbb{Q}$,

$$\mathrm{d}Y_t = -(a_t + b_t Y_t)\mathrm{d}t + Z_t \cdot \mathrm{d}B_t,$$

we have reduced to the case where the generator does not depend on z. We next get rid of the linear term in y by introducing

$$\overline{Y}_t := Y_t e^{\int_0^t b_s \mathrm{d}s} \text{ so that } \mathrm{d}\overline{Y}_t = -a_t e^{\int_0^t b_s \mathrm{d}s}\mathrm{d}t + Z_t e^{\int_0^t b_s \mathrm{d}s}\mathrm{d}B_t.$$

Finally, defining

$$\overline{\overline{Y}}_t := \overline{Y}_t + \int_0^t a_u e^{\int_0^u b_s \mathrm{d}s}\mathrm{d}u,$$

we arrive at a BSDE with zero generator for $\overline{\overline{Y}}_t$ which can be solved by the martingale representation theorem under the equivalent probability measure $\mathbb{Q}$.

Of course, one can also express the solution under $\mathbb{P}$:

$$Y_t = \mathbb{E}\left[\Gamma_T^t \xi + \int_t^T \Gamma_s^t a_s \mathrm{d}s \Big| \mathcal{F}_t\right], \ t \le T,$$

where

$$\Gamma_s^t := \exp\left(\int_t^s b_u \mathrm{d}u - \frac{1}{2}\int_t^s |c_u|^2 \mathrm{d}u + \int_t^s c_u \cdot \mathrm{d}W_u\right), \quad 0 \le t \le s \le T. \quad (10.12)$$

10.2.3 The Main Existence and Uniqueness Result

The following result was proved by Pardoux and Peng [32].

Theorem 10.2. *Assume that* $\{f_t(0,0), t \in [0,T]\} \in \mathbb{H}^2$ *and, for some constant* $C > 0$,

$$\left|f_t(y,z) - f_t(y',z')\right| \le C(|y-y'| + |z-z'|), \ \mathrm{d}t \otimes \mathrm{d}\mathbb{P} - a.s.$$

for all $t \in [0,T]$ *and* $(y,z),(y',z') \in \mathbb{R}^n \times \mathbb{R}^{n\times d}$. *Then, for every* $\xi \in \mathbb{L}^2$, *there is a unique solution* $(Y,Z) \in \mathcal{S}^2 \times \mathbb{H}^2$ *to the BSDE*(f,ξ).

Proof. Denote $S = (Y,Z)$, and introduce the equivalent norm in the corresponding $\mathbb{H}^2$ space:

$$\|S\|_\alpha := \mathbb{E}\left[\int_0^T \mathrm{e}^{\alpha t}(|Y_t|^2 + |Z_t|^2)\mathrm{d}t\right],$$

where α will be fixed later. We consider the operator

$$\phi : s = (y,z) \in \mathbb{H}^2 \longmapsto S^s = (Y^s, Z^s)$$

defined by:

$$Y_t^s = \xi + \int_t^T f_u(y_u,z_u)\mathrm{d}u - \int_t^T Z_u^s \cdot \mathrm{d}W_u, \ \ t \le T.$$

1. First, since $|f_u(y_u,z_u)| \le |f_u(0,0)| + C(|y_u| + |z_u|)$, we see that the process $\{f_u(y_u,z_u), u \le T\}$ is in $\mathbb{H}^2$. Then S^s is well defined by the martingale representation theorem and $S^s = \phi(s) \in \mathbb{H}^2$.
2. For $s,s' \in \mathbb{H}^2$, denote $\delta s := s - s'$, $\delta S := S^s - S^{s'}$ and $\delta f := f_t(S^s) - f_t(S^{s'})$. Since $\delta Y_T = 0$, it follows from Itô's formula that

$$e^{\alpha t}|\delta Y_t|^2 + \int_t^T e^{\alpha u}|\delta Z_u|^2 du = \int_t^T e^{\alpha u}\left(2\delta Y_u \cdot \delta f_u - \alpha|\delta Y_u|^2\right) du$$
$$-2\int_t^T e^{\alpha u}(\delta Z_u)^{\mathrm{T}}\delta Y_u \cdot dW_u.$$

In the remaining part of this step, we prove that

$$M_. := \int_0^{\cdot} e^{\alpha u}(\delta Z_u)^{\mathrm{T}}\delta Y_u \cdot dW_u \text{ is a uniformly integrable martingale,} \quad (10.13)$$

so that we deduce from the previous equality that

$$\mathbb{E}\left[e^{\alpha t}|\delta Y_t|^2 + \int_t^T e^{\alpha u}|\delta Z_u|^2 du\right] = \mathbb{E}\left[\int_t^T e^{\alpha u}\left(2\delta Y_u \cdot \delta f_u - \alpha|\delta Y_u|^2\right) du\right]. \quad (10.14)$$

To prove (10.13), we verify that $\sup_{t\le T}|M_t| \in \mathbb{L}^1$. Indeed, by the Burkholder-Davis-Gundy inequality, we have

$$\mathbb{E}\left[\sup_{t\le T}|M_t|\right] \le C\mathbb{E}\left[\left(\int_0^T e^{2\alpha u}|\delta Y_u|^2|\delta Z_u|^2 du\right)^{1/2}\right]$$
$$\le C'\mathbb{E}\left[\sup_{u\le T}|\delta Y_u|\left(\int_0^T |\delta Z_u|^2 du\right)^{1/2}\right]$$
$$\le \frac{C'}{2}\left(\mathbb{E}\left[\sup_{u\le T}|\delta Y_u|^2\right] + \mathbb{E}\left[\int_0^T |\delta Z_u|^2 du\right]\right) < \infty.$$

3. We now continue estimating (10.14) by using the Lipschitz property of the generator:

$$\mathbb{E}\left[e^{\alpha t}|\delta Y_t|^2 + \int_t^T e^{\alpha u}|\delta Z_u|^2 du\right]$$
$$\le \mathbb{E}\left[\int_t^T e^{\alpha u}\left(-\alpha|\delta Y_u|^2 + C2|\delta Y_u|(|\delta y_u| + |\delta z_u|)\right) du\right]$$
$$\le \mathbb{E}\left[\int_t^T e^{\alpha u}\left(-\alpha|\delta Y_u|^2 + C\left(\varepsilon^2|\delta Y_u|^2 + \varepsilon^{-2}(|\delta y_u| + |\delta z_u|)^2\right)\right) du\right]$$

for any $\varepsilon > 0$. Choosing $C\varepsilon^2 = \alpha$, we obtain:

$$\mathbb{E}\left[e^{\alpha t}|\delta Y_t|^2 + \int_t^T e^{\alpha u}|\delta Z_u|^2 du\right] \le \mathbb{E}\left[\int_t^T e^{\alpha u}\frac{C^2}{\alpha}(|\delta y_u| + |\delta z_u|)^2 du\right]$$
$$\le 2\frac{C^2}{\alpha}\|\delta s\|_\alpha^2.$$

This provides

$$\|\delta Z\|_\alpha^2 \leq 2\frac{C^2}{\alpha}\|\delta s\|_\alpha^2 \text{ and } \|\delta Y\|_\alpha^2 \mathrm{d}t \leq 2\frac{C^2 T}{\alpha}\|\delta s\|_\alpha^2$$

where we abused notation by writing $\|\delta Y\|_\alpha$ and $\|\delta Z\|_\alpha$ although these processes do not have the dimension required by the definition. Finally, these two estimates imply that

$$\|\delta S\|_\alpha \leq \sqrt{\frac{2C^2}{\alpha}(1+T)}\|\delta s\|_\alpha.$$

By choosing $\alpha > 2(1+T)C^2$, it follows that the map ϕ is a contraction on $\mathbb{H}^2$, and that there is a unique fixed point.

4. It remain, to prove that $Y \in \mathcal{S}^2$. This is easily obtained by first estimating:

$$\mathbb{E}\left[\sup_{t\leq T}|Y_t|^2\right] \leq C\left(|Y_0|^2 + \mathbb{E}\left[\int_0^T |f_t(Y_t, Z_t)|^2 \mathrm{d}t\right]\right.$$
$$\left. + \mathbb{E}\left[\sup_{t\leq T}\left|\int_0^t Z_s \cdot \mathrm{d}W_s\right|^2\right]\right),$$

and then using the Lipschitz property of the generator and the Burkholder-Davis-Gundy inequality. □

Remark 10.3. Consider the Picard iterations:

$$(Y^0, Z^0) = (0, 0), \text{ and}$$
$$Y_t^{k+1} = \xi + \int_t^T f_s(Y_s^k, Z_s^k)\mathrm{d}s + \int_t^T Z_s^{k+1} \cdot \mathrm{d}W_s.$$

Given (Y^k, Z^k), the next step (Y^{k+1}, Z^{k+1}) is defined by means of the martingale representation theorem. Then, $S^k = (Y^k, Z^k) \longrightarrow (Y, Z)$ in $\mathbb{H}^2$ as $k \to \infty$. Moreover, since

$$\|S^k\|_\alpha \leq \left(\frac{2C^2}{\alpha}(1+T)\right)^k,$$

it follows that $\sum_k \|S^k\|_\alpha < \infty$, and we conclude by the Borel–Cantelli lemma that the convergence $(Y^k, Z^k) \longrightarrow (Y, Z)$ also holds $\mathrm{d}t \otimes \mathrm{d}\mathbb{P}$–a.s.

10.3 Comparison and Stability

Theorem 10.4. *Let $n = 1$, and let (Y^i, Z^i) be the solution of BSDE(f^i, ξ^i) for some pair (ξ^i, f^i) satisfying the conditions of Theorem 10.2, $i = 0, 1$. Assume that*

$$\xi^1 \geq \xi^0 \text{ and } f_t^1(Y_t^0, Z_t^0) \geq f_t^0(Y_t^0, Z_t^0),\ \mathrm{d}t \otimes \mathrm{d}\mathbb{P} - a.s. \tag{10.15}$$

Then $Y_t^1 \geq Y_t^0$, $t \in [0, T]$, $\mathbb{P}$–a.s.

Proof. We denote

$$\delta Y := Y^1 - Y^0,\ \delta Z := Z^1 - Z^0,\ \delta_0 f := f^1(Y^0, Z^0) - f^0(Y^0, Z^0),$$

and we compute that

$$\mathrm{d}(\delta Y_t) = -(\alpha_t \delta Y_t + \beta_t \cdot \delta Z_t + \delta_0 f_t)\,\mathrm{d}t + \delta Z_t \cdot \mathrm{d}W_t, \tag{10.16}$$

where

$$\alpha_t := \frac{f_t^1(Y_t^1, Z_t^1) - f_t^1(Y_t^0, Z_t^1)}{\delta Y_t} \mathbf{1}_{\{\delta Y_t \neq 0\}},$$

and, for $j = 1, \dots, d$,

$$\beta_t^j := \frac{f_t^1\left(Y_t^0, Z_t^1 \oplus_{j-1} Z_t^0\right) - f_t^1\left(Y_t^0, Z_t^1 \oplus_j Z_t^0\right)}{\delta Z_t^{0,j}} \mathbf{1}_{\{\delta Z_t^{0,j} \neq 0\}},$$

where $\delta Z^{0,j}$ denotes the j-th component of δZ^0, and for every $z^0, z^1 \in \mathbb{R}^d$, $z^1 \oplus_j z^0 := \left(z^{1,1}, \dots, z^{1,j}, z^{0,j+1}, \dots, z^{0,d}\right)$ for $0 < j < d$, $z^1 \oplus_0 z^0 := z^0$, $z^1 \oplus_d z^0 := z^1$.

Since f^1 is Lipschitz-continuous, the processes α and β are bounded. Solving the linear BSDE (10.16) as in Sect. 10.2.2, we get

$$\delta Y_t = \mathbb{E}\left[\Gamma_T^t \delta Y_T + \int_t^T \Gamma_u^t \delta_0 f_u \mathrm{d}u \middle| \mathcal{F}_t\right],\ t \leq T,$$

where the process Γ^t is defined as in (10.12) with $(\delta_0 f, \alpha, \beta)$ substituted to (a, b, c). Then Condition (10.15) implies that $\delta Y \geq 0$, $\mathbb{P}$–a.s. □

Our next result compares the difference in absolute value between the solutions of the two BSDEs and provides a bound which depends on the difference between the corresponding final datum and the generators. In particular, this bound provides a transparent information about the nature of conditions needed to pass to limits with BSDEs.

Theorem 10.5. *Let (Y^i, Z^i) be the solution of BSDE(f^i, ξ^i) for some pair (f^i, ξ^i) satisfying the conditions of Theorem 10.2, $i = 0, 1$. Then*

$$\|Y^1 - Y^0\|_{\mathcal{S}^2}^2 + \|Z^1 - Z^0\|_{\mathbb{H}^2}^2 \le C\left(\|\xi^1 - \xi^0\|_{\mathbb{L}^2}^2 + \|(f^1 - f^0)(Y^0, Z^0)\|_{\mathbb{H}^2}^2\right),$$

where C is a constant depending only on T and the Lipschitz constant of f^1.

Proof. We denote $\delta\xi := \xi^1 - \xi^0$, $\delta Y := Y^1 - Y^0$, $\delta f := f^1(Y^1, Z^1) - f^0(Y^0, Z^0)$, and $\Delta f := f^1 - f^0$. Given a constant β to be fixed later, we compute by Itô's formula that

$$\mathrm{e}^{\beta t}|\delta Y_t|^2 = \mathrm{e}^{\beta T}|\delta\xi|^2 + \int_t^T \mathrm{e}^{\beta u}\left(2\delta Y_u \cdot \delta f_u - |\delta Z_u|^2 - \beta|\delta Y_u|^2\right)\mathrm{d}u$$
$$+2\int_t^T \mathrm{e}^{\beta u}\delta Z_u^{\mathrm{T}}\delta Y_u \cdot \mathrm{d}W_u.$$

By the same argument as in the proof of Theorem 10.2, we see that the stochastic integral term has zero expectation. Then

$$\mathrm{e}^{\beta t}|\delta Y_t|^2 = \mathbb{E}_t\left[\mathrm{e}^{\beta T}|\delta\xi|^2 + \int_t^T \mathrm{e}^{\beta u}\left(2\delta Y_u \cdot \delta f_u - |\delta Z_u|^2 - \beta|\delta Y_u|^2\right)\mathrm{d}u\right], \tag{10.17}$$

where $\mathbb{E}_t := \mathbb{E}[.|\mathcal{F}_t]$. We now estimate that, for any $\varepsilon > 0$,

$$\begin{aligned}
2\delta Y_u \cdot \delta f_u &\le \varepsilon^{-1}|\delta Y_u|^2 + \varepsilon|\delta f_u|^2 \\
&\le \varepsilon^{-1}|\delta Y_u|^2 + \varepsilon\big(C(|\delta Y_u| + |\delta Z_u|) + |\Delta f_u(Y_u^0, Z_u^0)|\big)^2 \\
&\le \varepsilon^{-1}|\delta Y_u|^2 + 3\varepsilon\big(C^2(|\delta Y_u|^2 + |\delta Z_u|^2) + |\Delta f_u(Y_n^0, Z_u^0)|^2\big).
\end{aligned}$$

We then choose $\varepsilon := 1/(6C^2)$ and $\beta := 3\varepsilon C^2 + \varepsilon^{-1}$, and plug the latter estimate in (10.17). This provides:

$$\mathrm{e}^{\beta t}|\delta Y_t|^2 + \frac{1}{2}\mathbb{E}_t\left[\int_t^T |\delta Z_u|^2 \mathrm{d}u\right] \le \mathbb{E}_t\left[\mathrm{e}^{\beta T}|\delta\xi|^2 + \frac{1}{2C^2}\int_0^T \mathrm{e}^{\beta u}|\Delta f_u(Y_u^0, Z_u^0)|^2 \mathrm{d}u\right],$$

which implies the required inequality by taking the supremum over $t \in [0, T]$ and using the Doob's maximal inequality for the martingale $\{\mathbb{E}_t[\mathrm{e}^{\beta T}|\delta\xi|^2], t \le T\}$. □

10.4 BSDEs and Stochastic Control

We now turn to the question of controlling the solution of a family of BSDEs in the scalar case $n = 1$. Let $(f_\nu, \xi_\nu)_{\nu\in\mathcal{U}}$ be a family of coefficients, where $\mathcal{U}$ is some given set of controls. We assume that the coefficients $(f_\nu, \xi_\nu)_{\nu\in\mathcal{U}}$ satisfy the conditions of the existence and uniqueness of Theorem 10.2, and we consider the following stochastic control problem:

$$V_0 := \sup_{\nu \in \mathcal{U}} Y_0^\nu, \tag{10.18}$$

where (Y^ν, Z^ν) is the solution of BSDE(f^ν, ξ^ν).

The above stochastic control problem boils down to the standard control problems of Sect. 3.1 when the generators f^α are all zero. When the generators f^ν are affine in (y, z), the problem (10.18) can also be recasted in the standard framework, by discounting and change of measure.

The following easy result shows that the above maximization problem can be solved by maximizing the coefficients (ξ^α, f^α):

$$f_t(y,z) := \operatorname{ess\,sup}_{\nu \in \mathcal{U}} f_t^\nu(y,z), \quad \xi := \operatorname{ess\,sup}_{\nu \in \mathcal{U}} \xi^\nu. \tag{10.19}$$

The notion of essential supremum is recalled in the Appendix of this chapter. We will assume that the coefficients (f, ξ) satisfy the conditions of the existence result of Theorem 10.2, and we will denote by (Y, Z) the corresponding solution.

A careful examination of the statement below shows a great similarity with the verification result in stochastic control. In the present non-Markovian framework, this remarkable observation shows that the notion of BSDEs allows to mimic the stochastic control methods developed in the Markovian framework of the previous chapters.

Proposition 10.6. *Assume that the coefficients (f, ξ) and (f_ν, ξ_ν) satisfy the conditions of Theorem 10.2, for all $\nu \in \mathcal{U}$. Assume further that there exists some $\hat{\nu} \in \mathcal{U}$ such that*

$$f_t(y,z) = f^{\hat{\nu}}(y,z) \text{ and } \xi = \xi^{\hat{\nu}}.$$

Then $V_0 = Y_0^{\hat{\nu}}$ and $Y_t = \operatorname{ess\,sup}_{\nu \in \mathcal{U}} Y_t^\nu$, $t \in [0, T]$, $\mathbb{P}-a.s.$

Proof. The $\mathbb{P}-$a.s. inequality $Y \geq Y^\nu$, for all $\nu \in \mathcal{U}$, is a direct consequence of the comparison result of Theorem 10.4. Hence, $Y_t \geq \sup_{\nu \in \mathcal{U}} Y_t^\nu$, $\mathbb{P}-$a.s. To conclude, we notice that Y and $Y^{\hat{\nu}}$ are two solutions of the same BSDE, and therefore must coincide, by uniqueness. □

The next result characterizes the solution of a standard stochastic control problem in terms of a BSDE. Here, again, we emphasize that, in the present non-Markovian framework, the BSDE is playing the role of the dynamic programming equation whose scope is restricted to the Markovian case.

Let

$$U_0 := \inf_{\nu \in \mathcal{U}} \mathbb{E}^{\mathbb{P}^\nu}\left[\beta_{0,T}^\nu \xi^\nu + \int_0^T \beta_{u,T}^\nu \ell_u(\nu_u) du\right],$$

where

$$\left.\frac{\mathrm{d}\mathbb{P}^\nu}{\mathrm{d}\mathbb{P}}\right|_{\mathcal{F}_T} := \mathrm{e}^{\int_0^T \lambda_t(\nu_t)\cdot \mathrm{d}W_t - \frac{1}{2}\int_0^T |\lambda_t(\nu_t)|^2 \mathrm{d}t} \text{ and } \beta^\nu_{t,T} := \mathrm{e}^{-\int_t^T k_u(\nu_u)\mathrm{d}u}.$$

We assume that all coefficients involved in the above expression satisfy the required conditions for the problem to be well defined.

We first notice that for every $\nu \in \mathcal{U}$, the process

$$Y^\nu_t := \mathbb{E}^{\mathbb{P}^\nu}\left[\beta^\nu_{t,T}\xi^\nu + \int_t^T \beta^\nu_{u,T}\ell_u(\nu_u)\mathrm{d}u \Big| \mathcal{F}_t\right], \quad t \in [0,T],$$

is the first component of the solution (Y^ν, Z^ν) of the affine BSDE:

$$\mathrm{d}Y^\nu_t = -f^\nu_t(Y^\nu_t, Z^\nu_t)\mathrm{d}t + Z^\nu_t \mathrm{d}W_t, \quad Y^\nu_T = \xi^\nu,$$

with $f^\nu_t(y,z) := \ell_t(\nu_t) - k_t(\nu_t)y + \lambda_t(\nu_t)z$. In view of this observation, the following result is a direct application of Proposition 10.6.

Proposition 10.7. *Assume that the coefficients*

$$\xi := \operatorname*{ess\,sup}_{\nu\in\mathcal{U}} \xi^\nu \quad \text{and} \quad f_t(y,z) := \operatorname*{ess\,sup}_{\nu\in\mathcal{U}} f^\nu_t(y,z)$$

satisfy the conditions of Theorem 10.2, and let (Y, Z) be the corresponding solution. Then $U_0 = Y_0$.

10.5 BSDEs and Semilinear PDEs

In this section, we specialize the discussion to the so-called Markovian BSDEs in the one-dimensional case $n = 1$. This class of BSDEs corresponds to the case where

$$f_t(y,z) = F(t, X_t, y, z) \text{ and } \xi = g(X_T),$$

where $F : [0,T] \times \mathbb{R}^d \times \mathbb{R} \times \mathbb{R}^d \longrightarrow \mathbb{R}$ and $g : \mathbb{R}^d \longrightarrow \mathbb{R}$ are measurable, and X is a Markovian diffusion process defined by some initial data X_0 and the SDE:

$$\mathrm{d}X_t = \mu(t, X_t)\mathrm{d}t + \sigma(t, X_t)\mathrm{d}W_t. \tag{10.20}$$

Here μ and σ are continuous and satisfy the usual Lipschitz and linear growth conditions in order to ensure existence and uniqueness of a strong solution to the SDE (10.20), and

$$F, g \text{ have polynomial growth in x,}$$
$$\text{and } F \text{ is uniformly Lipschitz in } (y,z).$$

Then, it follows from Theorem 10.2 that the above Markovian BSDE has a unique solution.

We next move the time origin by considering the solution $\{X_s^{t,x}, s \geq t\}$ of (10.20) with initial data $X_t^{t,x} = x$. The corresponding solution of the BSDE

$$dY_s = -F(s, X_s^{t,x}, Y_s, Z_s)ds + Z_s dW_s, \quad Y_T = g\left(X_T^{t,x}\right) \tag{10.21}$$

will be denoted by $(Y^{t,x}, Z^{t,x})$.

Proposition 10.8. *The process $\{\left(Y_s^{t,x}, Z_s^{t,x}\right), s \in [t,T]\}$ is adapted to the filtration*

$$\mathcal{F}_s^t := \sigma\left(W_u - W_t, u \in [t,s]\right), \quad s \in [t,T].$$

In particular, $u(t,x) := Y_t^{t,x}$ is a deterministic function and

$$Y_s^{t,x} \;=\; Y_s^{s,X_s^{t,x}} = u\left(s, X_s^{t,x}\right), \quad \textit{for all } s \in [t,T], \; \mathbb{P}-a.s.$$

Proof. The first claim is obvious, and the second one follows from the fact that $X_r^{t,x} = X_r^{s,X_s^{t,x}}$. □

Proposition 10.9. *Let u be the function defined in Proposition 10.8, and assume that $u \in C^{1,2}([0,T), \mathbb{R}^d)$. Then*

$$-\partial_t u - \mu \cdot Du - \frac{1}{2} Tr[\sigma\sigma^T D^2 u] - f(., u, \sigma^T Du) = 0 \textit{ on } [0,T) \times \mathbb{R}^d. \tag{10.22}$$

Proof. This an easy application of Itô's formula together with the usual localization technique. □

By weakening the interpretation of the PDE (10.22) to the sense of viscosity solutions, we may drop the regularity condition on the function u in the latter statement. We formulate this result in the following exercise.

Exercise 10.10. Show that the function u of Proposition 10.8 is a viscosity solution of the semilinear PDE (10.22), i.e., u_* and u^* are viscosity supersolution and subsolutions of (10.22), respectively.

We conclude this chapter by nonlinear version of the Feynman–Kac formula.

Theorem 10.11. *Let $v \in C^{1,2}([0,T), \mathbb{R}^d)$ be a solution of the semilinear PDE (10.22) with polynomially growing v and $\sigma^T Dv$. Then*

$$v(t,x) = Y_t^{t,x} \quad \textit{for all } (t,x) \in [0,T] \times \mathbb{R}^d,$$

where $(Y^{t,x},\ Z^{t,x})$ is the solution of the BSDE (10.21).

Proof. For fixed (t,x), denote $Y_s := v(s, X_s^{t,x})$ and $Z_s := \sigma^T(s, X_s^{t,x})$. Then, it follows from Itô's formula that (Y,Z) solves (10.21). From the polynomial growth on v and Dv, we see that the processes Y and Z are both in $\mathbb{H}^2$. Then they coincide with the unique solution of (10.21). □

10.6 Appendix: Essential Supremum

The notion of essential supremum has been introduced in probability in order to face the problem of maximizing random variables over an infinite family $\mathcal{Z}$. The problem arises when $\mathcal{Z}$ is not countable because then the supremum is not measurable, in general.

Theorem 10.12. *Let $\mathcal{Z}$ be a family of r.v. $Z : \Omega \longrightarrow \mathbb{R} \cup \{\infty\}$ on a probability space $(\Omega, \mathcal{F}, \mathbb{P})$. Then there exists a unique (a.s.) r.v. $\bar{Z} : \Omega \longrightarrow \mathbb{R} \cup \{\infty\}$ such that*

(a) $\bar{Z} \geq Z$*, a.s. for all* $Z \in \mathcal{Z}$.
(b) *For all r.v.* Z' *satisfying* (a), *we have* $\bar{Z} \leq Z'$*, a.s.*

Moreover, there exists a sequence $(Z_n)_{n \in \mathbb{N}} \subset \mathcal{Z}$ such that $\bar{Z} = \sup_{n \in \mathbb{N}} Z_n$.
The r.v. $\bar{Z}$ is called the essential supremum of the family $\mathcal{Z}$ and denoted by ess sup $\mathcal{Z}$.

Proof. The uniqueness of $\bar{Z}$ is an immediate consequence of (b). To prove existence, we consider the set $\mathcal{D}$ of all countable subsets of $\mathcal{Z}$. For all $D \in \mathcal{D}$, we define $Z_D := \sup\{Z : Z \in D\}$, and we introduce the r.v. $\zeta := \sup\{\mathbb{E}[Z_D] : D \in \mathcal{D}\}$.

1. We first prove that there exists $D^* \in \mathcal{D}$ such that $\zeta = \mathbb{E}[Z_{D^*}]$. To see this, let $(D_n)_n \subset \mathcal{D}$ be a maximizing sequence, i.e., $\mathbb{E}[Z_{D_n}] \longrightarrow \zeta$, then $D^* := \cup_n D_n \in \mathcal{D}$ satisfies $\mathbb{E}[Z_{D^*}] = \zeta$. We denote $\bar{Z} := Z_{D^*}$.
2. It is clear that the r.v. $\bar{Z}$ satisfies (b). To prove that property (a) holds true, we consider an arbitrary $Z \in \mathcal{Z}$ together with the countable family $D := D^* \cup \{Z\} \subset \mathcal{D}$. Then $Z_D = Z \vee \bar{Z}$, and $\zeta = \mathbb{E}[\bar{Z}] \leq \mathbb{E}[Z \vee \bar{Z}] \leq \zeta$. Consequently, $Z \vee \bar{Z} = \bar{Z}$, and $Z \leq \bar{Z}$, a.s. □

Chapter 11
Quadratic Backward SDEs

In this chapter, we consider an extension of the notion of BSDEs to the case where the dependence of the generator in the variable z has quadratic growth. In the Markovian case, this corresponds to a problem of second-order semilinear PDE with quadratic growth in the gradient term. The first existence and uniqueness result in this context was established by M. Kobylanski in her Ph.D. thesis by adapting some previously established PDE techniques to the non-Markovian BSDE framework. In this chapter, we present an alternative argument introduced recently by Tevzadze [39].

Quadratic BSDEs turn out to play an important role in the applications, and the extension of this section is needed in order to analyze the problem of portfolio optimization under portfolio constraints.

We shall consider throughout this chapter the BSDE

$$Y_t = \xi + \int_t^T f_s(Z_s)\mathrm{d}s - \int_t^T Z_s \cdot \mathrm{d}W_s \qquad (11.1)$$

where ξ is a bounded $\mathcal{F}_T$–measurable r.v. and $f : [0,T] \times \Omega \times \mathbb{R}^d \longrightarrow \mathbb{R}$ is $\mathcal{P} \otimes \mathcal{B}(\mathbb{R}^d)$–measurable and satisfies a quadratic growth condition:

$$\|\xi\|_\infty < \infty \text{ and } |f_t(z)| \le C(1 + |z|^2) \text{ for some constant } C > 0. \qquad (11.2)$$

We could have included a Lipschitz dependence of the generator on the variable y without altering the results of this chapter. However, for exposition clarity and transparency, we drop this dependence in order to concentrate on the main difficulty, namely, the quadratic growth in z.

N. Touzi, *Optimal Stochastic Control, Stochastic Target Problems, and Backward SDE*, Fields Institute Monographs 29, DOI 10.1007/978-1-4614-4286-8_11,

11.1 A Priori Estimates and Uniqueness

In this section, we prove two easy results. First, we show the connection between the boundedness of the component Y of the solution and the *bounded mean oscillation* (BMO) property for the martingale part $\int_0^\cdot Z_t \cdot \mathrm{d}W_t$. Then, we prove uniqueness in this class.

11.1.1 A Priori Estimates for Bounded Y

We denote by $\mathcal{M}^2$ the collection of all $\mathbb{P}$−square integrable martingales on the time interval $[0, T]$. We first introduce the so-called class of martingales with BMO:

$$\mathrm{BMO} := \{M \in \mathcal{M}^2 : \|M\|_{\mathrm{BMO}} < \infty\},$$

where

$$\|M\|_{\mathrm{BMO}} := \sup_{\tau \in \mathcal{T}_0^T} \|\mathbb{E}[\langle M\rangle_T - \langle M\rangle_\tau | \mathcal{F}_\tau]\|_\infty .$$

Here, $\mathcal{T}_0^T$ is the collection of all stopping times, and $\langle M\rangle$ denotes the quadratic variation process of M. We will be essentially working with square integrable martingales of the form $M = \int_0^\cdot \phi_s \mathrm{d}W_s$. The following definition introduces an abuse of notation which will be convenient for our presentation.

Definition 11.1. A process $\phi \in \mathbb{H}^2$ is said to be a BMO martingale generator if

$$\|\phi\|_{\mathbb{H}^2_{\mathrm{BMO}}} := \left\| \int_0^\cdot \phi_s \cdot \mathrm{d}W_s \right\|_{\mathrm{BMO}} < \infty.$$

We denote by $\mathbb{H}^2_{\mathrm{BMO}} := \{\phi \in \mathbb{H}^2 : \|\phi\|_{\mathbb{H}^2_{\mathrm{BMO}}} < \infty\}$.

For this class of martingales, we can rewrite the BMO norm by the Itô isometry into

$$\|\phi\|^2_{\mathbb{H}^2_{\mathrm{BMO}}} := \sup_{\tau \in \mathcal{T}_0^T} \left\| \mathbb{E}\left[\int_\tau^T |\phi_s|^2 \mathrm{d}s \big| \mathcal{F}_\tau\right]\right\|_\infty .$$

The following result shows why this notion is important in the context of quadratic BSDEs.

Lemma 11.2. *Let* (Y, Z) *be a solution of the quadratic BSDE* (11.1) *(in particular,* $Z \in \mathbb{H}^2_{loc}$*) with generator* f *satisfying* (11.2)*. Assume that the process* Y *is bounded. Then* $Z \in \mathbb{H}^2_{\mathrm{BMO}}$.

Proof. Let $(\tau_n)_{n\geq 1} \subset \mathcal{T}_0^T$ be a localizing sequence of the local martingale $\int_0^\cdot Z_s \cdot \mathrm{d}W_s$. By Itô's formula together with the boundedness of Y, we have for any $\tau \in \mathcal{T}_0^T$:

$$e^{\beta\|Y\|_\infty} \geq e^{\beta Y_{\tau_n}} - e^{\beta Y_\tau} = \int_\tau^{\tau_n} \beta e^{\beta Y_s} \left(\left(\frac{1}{2}\beta |Z_s|^2 - f_s(Z_s) \right) ds + Z_s \cdot dW_s \right).$$

By the Doob's optional sampling theorem, this provides

$$\begin{aligned}
\frac{\beta^2}{2}\mathbb{E}\left[\int_\tau^{\tau_n} e^{\beta Y_s} |Z_s|^2 ds \Big| \mathcal{F}_\tau \right] &\leq e^{\beta\|Y\|_\infty} + \beta \mathbb{E}\left[\int_\tau^{\tau_n} e^{\beta Y_s} f_s(Z_s) ds \Big| \mathcal{F}_\tau \right] \\
&\leq (1 + \beta C T) e^{\beta\|Y\|_\infty} + \beta C \mathbb{E}\left[\int_\tau^{\tau_n} e^{\beta Y_s} |Z_s|^2 ds \Big| \mathcal{F}_\tau \right].
\end{aligned}$$

Then, setting $\beta = 4C$, it follows that

$$\begin{aligned}
e^{-\beta\|Y\|_\infty} \mathbb{E}\left[\int_\tau^T |Z_s|^2 ds \Big| \mathcal{F}_\tau \right] &\leq \mathbb{E}\left[\int_\tau^T e^{\beta Y_s} |Z_s|^2 ds \Big| \mathcal{F}_\tau \right] \\
&= \lim_{n\to\infty} \uparrow \mathbb{E}\left[\int_\tau^{\tau_n} e^{\beta Y_s} |Z_s|^2 ds \Big| \mathcal{F}_\tau \right] \\
&\leq \frac{1 + 4C^2 T}{4C^2} e^{\beta\|Y\|_\infty},
\end{aligned}$$

which provides the required result by the arbitrariness of $\tau \in \mathcal{T}_0^T$. □

11.1.2 Some Properties of BMO Martingales

In this section, we list without proof some properties of the space BMO. We refer to the book of Kazamaki [25] for a complete presentation on this topic.

1. The set BMO is a Banach space.
2. $M \in$ BMO if and only if $\int H dM \in$ BMO for all bounded progressively measurable process H.
3. If $M \in$ BMO, then
   (a) The process $\mathcal{E}(M) := e^{M - \frac{1}{2}\langle M \rangle}$ is a uniformly integrable martingale.
   (b) The process $M - \langle M \rangle$ is a BMO martingale under the equivalent measure $\mathcal{E}(M) \cdot \mathbb{P}$.
   (c) $\mathcal{E}(M) \in \mathbb{L}^r$ for some $r > 1$.
4. For $\phi \in \mathbb{H}^2_{\text{BMO}}$, we have

$$\mathbb{E}\left[\left(\int_0^T |\phi_s|^2 ds \right)^p \right] \leq 2p! \left(4\|\phi\|^2_{\mathbb{H}^2_{\text{BMO}}} \right)^p \quad \text{for all } p \geq 1.$$

In our subsequent analysis, we shall only make use of the Properties 1 and 3.

11.1.3 Uniqueness

We now introduce the main condition for the derivation of the existence and uniqueness result.

Assumption 11.3. *The quadratic generator f is C^2 in z, and there are constants θ_1, θ_2 such that*

$$|D_z f_t(z)| \leq \theta_1(1+|z|),\ |D^2_{zz} f_t(z)| \leq \theta_2 \quad \text{for all } (t,\omega,z) \in [0,T]\times\Omega\times\mathbb{R}^d.$$

Lemma 11.4. *Let Assumption 11.3 hold true. Then, there exists a bounded progressively measurable process ϕ such that for all $t \in [0,T], z, z' \in \mathbb{R}^d$*

$$\left|f_t(z) - f_t(z') - \phi_t \cdot (z-z')\right| \leq \theta_2|z-z'|\left(|z|+|z'|\right),\ \mathbb{P}-a.s. \tag{11.3}$$

Proof. Since f is C^2 in z, we introduce the process $\phi_t := D_z f_t(0)$ which is bounded by θ_1, according to Assumption 11.3. By the mean value theorem, we compute that, for some constant $\lambda = \lambda(\omega) \in [0,1]$,

$$\begin{aligned}\left|f_t(z) - f_t(z') - \phi_t \cdot (z-z')\right| &= \left|D_z f_t(\lambda z + (1-\lambda)z') - \phi_t\right|\ |z-z'| \\ &\leq \theta_2|\lambda z + (1-\lambda)z'|\ |z-z'|,\end{aligned}$$

by the bound on $D^2_{zz} f_t(z)$ in Assumption 11.3. The required result follows from the trivial inequality $|\lambda z + (1-\lambda)z'| \leq |z| + |z'|$. □

We are now ready for the proof of the uniqueness result. As in the Lipschitz case, we have the following comparison result which implies uniqueness.

Theorem 11.5. *Let f^0, f^1 be two quadratic generators satisfying* (11.2). *Assume further that f^1 satisfies Assumption 11.3. Let (Y^i, Z^i), $i = 0, 1$, be two bounded solutions of* (11.1) *with coefficients (f^i, ξ^i). Assume that*

$$\xi^1 \geq \xi^0 \text{ and } f^1_t(Z^0_t) \geq f^0_t(Z^0_t),\ t \in [0,T],\ \mathbb{P}-a.s.$$

Then $Y^1 \geq Y^0$, $\mathbb{P}$–a.s.

Proof. We denote $\delta\xi := \xi^1 - \xi^0$, $\delta Y := Y^1 - Y^0$, $\delta Z := Z^1 - Z^0$, and $\delta f := f^1(Z^1) - f^0(Z^0)$. Then, it follows from Lemma 11.4 that

$$\begin{aligned}\delta Y_t &= \delta\xi - \int_t^T \delta Z_s \cdot dW_s + \int_t^T \delta f_s ds \\ &\geq \delta\xi - \int_t^T \delta Z_s \cdot dW_s + \int_t^T (f^1 - f^0)(Z^0_s) ds \\ &\quad + \int_t^T (f^1(Z^1_s) - f^1(Z^0_s)) ds\end{aligned}$$

$$
\begin{aligned}
&\geq \delta\xi - \int_t^T \delta Z_s \cdot \mathrm{d}W_s + \int_t^T (f^1 - f^0)(Z_s^0)\mathrm{d}s \\
&\quad + \int_t^T \Big(\phi_s \cdot (Z_s^1 - Z_s^0) - \theta_2 |Z_s^1 - Z_s^0|(|Z_s^0| + |Z_s^1|)\Big)\mathrm{d}s \\
&= \delta\xi - \int_t^T \delta Z_s \cdot (\mathrm{d}W_s - \Lambda_s \mathrm{d}s) + \int_t^T (f^1 - f^0)(Z_s^0)\mathrm{d}s,
\end{aligned}
$$

where ϕ is the bounded process introduced in Lemma 11.4 and the process Λ is defined by

$$
\Lambda_s := \phi_s - \theta_2 \frac{|Z_s^0| + |Z_s^1|}{|Z_s^1 - Z_s^0|}(Z_s^1 - Z_s^0)\mathbf{1}_{\{Z_s^1 - Z_s^0 \neq 0\}}, \ s \in [t, T].
$$

Since Y^0 and Y^1 are bounded, and both generators f^0, f^1 satisfy Condition (11.2), it follows from Lemma 11.2 that Z^0 and Z^1 are in $\mathbb{H}^2_{\mathrm{BMO}}$. Hence, $\Lambda \in \mathbb{H}^2_{\mathrm{BMO}}$, and by Property 3 of BMO martingales, we deduce that the process $W_. - \int_0^{\cdot} \Lambda_s \mathrm{d}s$ is a Brownian motion under an equivalent probability measure $\mathbb{Q}$. Taking conditional expectations under $\mathbb{Q}$ then provides

$$
\delta Y_t \geq \mathbb{E}_t^{\mathbb{Q}}\left[\delta\xi + \int_t^T (f^1 - f^0)(Z_s^0)\mathrm{d}s\right], \ \text{a.s.}
$$

which implies the required comparison result. □

11.2 Existence

In this section, we prove existence of a solution to the quadratic BSDE in two steps. We first prove existence (and uniqueness) by a fixed point argument when the final data ξ is bounded by some constant depending on the generator f and the maturity T. In the second step, we decompose the final data as $\xi = \sum_{i=1}^n \xi_i$ with ξ_i is sufficiently small so that the existence result of the first step applies. Then, we construct a solution of the quadratic BSDE with final data ξ by adding these solutions.

11.2.1 Existence for Small Final Condition

In this subsection, we prove an existence and uniqueness result for the quadratic BSDE (11.1) under Condition (11.3) with $\phi \equiv 0$. Recall that (11.3) was implied by Assumption 11.3.

Theorem 11.6. *Assume that the generator f satisfies*

$$f_t(0) = 0 \text{ and } \left|f_t(z) - f_t(z')\right| \leq \theta_2|z - z'|\left(|z| + |z'|\right), \ \mathbb{P} - a.s. \quad (11.4)$$

Then, for every $\mathcal{F}_T$−measurable r.v. ξ with $\|\xi\|_{\mathbb{L}^\infty} \leq (64\theta_2)^{-1}$, there exists a unique solution (Y, Z) to the quadratic BSDE (11.1) *with*

$$\|Y\|^2_{\mathcal{S}^\infty} + \|Z\|^2_{\mathbb{H}^2_{\mathrm{BMO}}} \leq (16\theta_2)^{-2}.$$

Proof. Consider the map $\Phi : (y, z) \in \mathcal{S}^\infty \times \mathbb{H}^2_{\mathrm{BMO}} \longmapsto S = (Y, Z)$ defined by:

$$Y_t = \xi + \int_t^T f_s(z_s)\mathrm{d}s - \int_t^T Z_s \cdot \mathrm{d}W_s, \ t \in [0, T], \ \mathbb{P} - \text{a.s.}$$

The existence of the pair $(Y, Z) = \Phi(y, z) \in \mathbb{H}^2$ is justified by the martingale representation theorem together with Property 4 of BMO martingales which ensures that the process $f(Z)$ is in $\mathbb{H}^2$.

To obtain the required result, we will prove that Φ is a contracting mapping on $\mathcal{S}^\infty \times \mathbb{H}^2_{\mathrm{BMO}}$ when ξ has a small $\mathbb{L}^\infty$−norm as in the statement of the theorem.

1. In this step, we prove that

$$(Y, Z) = \Phi(y, z) \in \mathcal{S}^\infty \times \mathbb{H}^2.$$

First, we estimate that

$$|Y_t| = \left|\mathbb{E}_t\left[\xi + \int_t^T f_s(z_s)\mathrm{d}s\right]\right|$$
$$\leq \|\xi\|_\infty + C\left(T + \|z\|_{\mathbb{H}^2_{\mathrm{BMO}}}\right),$$

proving that the process Y is bounded. We next calculate by Itô's formula that, for every stopping time $\tau \in \mathcal{T}_0^T$,

$$|Y_\tau|^2 + \mathbb{E}_\tau\left[\int_\tau^T |Z_s|^2\mathrm{d}s\right] = \mathbb{E}_\tau\left[|\xi|^2 + \int_\tau^T 2Y_s f_s(z_s)\mathrm{d}s\right]$$
$$\leq \|\xi\|^2_{\mathbb{L}^\infty} + 2\|Y\|_{\mathcal{S}^\infty}\mathbb{E}_\tau\left[\int_\tau^T |f_s(z_s)|\mathrm{d}s\right],$$

where $\mathbb{E}_\tau[.] = \mathbb{E}[.|\mathcal{F}_\tau]$ and, similar to the proof of Theorem 10.2, the expectation of the stochastic integral vanishes by Property 4 of BMO martingales.

By the trivial inequality $2ab \leq \frac{1}{4}a^2 + 4b^2$, it follows from the last inequality that:

$$|Y_\tau|^2 + \mathbb{E}_\tau\left[\int_\tau^T |Z_s|^2 ds\right] \leq \|\xi\|_{\mathbb{L}^\infty}^2 + \frac{1}{4}\|Y\|_{\mathcal{S}^\infty}^2 + 4\left(\mathbb{E}_\tau\left[\int_\tau^T |f_s(z_s)|ds\right]\right)^2$$
$$\leq \|\xi\|_{\mathbb{L}^\infty}^2 + \frac{1}{4}\|Y\|_{\mathcal{S}^\infty}^2 + 4\left(\mathbb{E}_\tau\left[\int_\tau^T \theta_2|z_s|^2 ds\right]\right)^2$$

by Condition (11.4). Taking the supremum over all stopping times $\tau \in \mathcal{T}_0^T$, this provides:

$$\|Y\|_{\mathcal{S}^\infty}^2 + \|Z\|_{\mathbb{H}^2_{\text{BMO}}}^2 \leq 2\|\xi\|_{\mathbb{L}^\infty}^2 + \frac{1}{2}\|Y\|_{\mathcal{S}^\infty}^2 + 8\theta_2^2\|z\|_{\mathbb{H}^2_{\text{BMO}}}^4,$$

and therefore

$$\|Y\|_{\mathcal{S}^\infty}^2 + \|Z\|_{\mathbb{H}^2_{\text{BMO}}}^2 \leq 4\|\xi\|_{\mathbb{L}^\infty}^2 + 16\theta_2^2\ \|z\|_{\mathbb{H}^2_{\text{BMO}}}^4.$$

The power 4 on the right-hand side is problematic as it may cause the explosion of the norms, given that the left-hand side is only raised to the power 2! This is precisely the reason why we need to restrict $\|\xi\|_{\mathbb{L}^\infty}$ to be small. For instance, let

$$R := \frac{1}{16\theta_2}, \quad \|\xi\|_{\mathbb{L}^\infty} \leq \frac{R}{4} \text{ and } \|y\|_{\mathcal{S}^\infty}^2 + \|z\|_{\mathbb{H}^2_{\text{BMO}}}^2 \leq R^2.$$

Then, it follows from the previous estimates that

$$\|Y\|_{\mathcal{S}^\infty}^2 + \|Z\|_{\mathbb{H}^2_{\text{BMO}}}^2 \leq 4\frac{R^2}{16} + 16\theta_2^2 R^4 = \frac{5R^2}{16}.$$

Denoting by B_R the ball of radius R in $\mathcal{S}^\infty \times \mathbb{H}^2_{\text{BMO}}$, we have then proved that

$$\Phi(B_R) \subset B_R.$$

2. For $i = 0, 1$ and $(y^i, z^i) \in B_R$, we denote $(Y^i, Z^i) := \Phi(y^i, z^i)$, $\delta y := y^1 - y^0$, $\delta z := z^1 - z^0$, $\delta Y := Y^1 - Y^0$, $\delta Z := Z^1 - Z^0$, and $\delta f := f(z^1) - f(z^0)$. We argue as in the previous step: apply Itô's formula for each stopping time $\tau \in \mathcal{T}_0^T$, take conditional expectations, and maximize over $\tau \in \mathcal{T}_0^T$. This leads to

$$\|\delta Y\|_{\mathcal{S}^\infty}^2 + \|\delta Z\|_{\mathbb{H}^2_{\text{BMO}}}^2 \leq 16 \sup_{\tau \in \mathcal{T}_0^T}\left(\mathbb{E}_\tau\left[\int_\tau^T |\delta f_s|ds\right]\right)^2. \qquad (11.5)$$

We next estimate that

$$\begin{aligned}\left(\mathbb{E}_\tau\left[\int_\tau^T |\delta f_s|\mathrm{d}s\right]\right)^2 &\le \theta_2^2\left(\mathbb{E}_\tau\left[\int_\tau^T |\delta z_s|(|z_s^0|+|z_s^1|)\mathrm{d}s\right]\right)^2\\ &\le \theta_2^2\,\mathbb{E}_\tau\left[\int_\tau^T |\delta z_s|^2\mathrm{d}s\right]\mathbb{E}_\tau\left[\int_\tau^T (|z_s^0|+|z_s^1|)^2\mathrm{d}s\right]\\ &\le 4R^2\theta_2^2\,\mathbb{E}_\tau\left[\int_\tau^T |\delta z_s|^2\mathrm{d}s\right].\end{aligned}$$

Then, it follows from (11.5) that

$$\|\delta Y\|_{\mathcal{S}^\infty}^2 + \|\delta Z\|_{\mathbb{H}^2_{\mathrm{BMO}}}^2 \le 16\times 4R^2\theta_2^2\,\|\delta z\|_{\mathbb{H}^2_{\mathrm{BMO}}}^2 \le \frac{1}{4}\,\|\delta z\|_{\mathbb{H}^2_{\mathrm{BMO}}}^2.$$

Hence, Φ is a contraction, and there is a unique fixed point. □

11.2.2 Existence for Bounded Final Condition

We now use the existence result of Theorem 11.6 to build a solution for a quadratic BSDE with general bounded final condition. Let us already observe that, in contrast with Theorem 11.6, the following construction will only provide existence (and not uniqueness) of a solution (Y, Z) with bounded Y component. However, this is all we need to prove in this section as the uniqueness is a consequence of Theorem 11.5.

We first observe that, under Condition (11.2), we may assume without loss of generality that $f_t(0) = 0$. This is an immediate consequence of the obvious equivalence:

$$(Y,Z)\text{ solution of BSDE}(f,\xi)\quad\text{iff } (\tilde{Y},Z)\text{ solution of BSDE}(f,\tilde{\xi}),$$

where $\tilde{Y}_t := Y_t - \int_0^t f_s(0)\mathrm{d}s$, $0 \le t \le T$, and $\tilde{\xi} := \xi - \int_0^T f_s(0)\mathrm{d}s$.

We then continue assuming that $f_t(0) = 0$.

Consider an arbitrary decomposition of the final data ξ as

$$\xi = \sum_{i=1}^n \xi_i \text{ where } \|\xi_i\|_{\mathbb{L}^\infty} \le \frac{1}{64\theta_2}. \tag{11.6}$$

For instance, one may simply take $\xi_i := \frac{1}{n}\xi$ and n sufficiently large so that (11.6) holds true.

We will then construct solutions (Y^i, Z^i) to quadratic BSDEs with final data ξ_i as follows:

Step 1. Let $f^1 := f$, and define (Y^1, Z^1) as the unique solution of the quadratic BSDE

$$Y_t^1 = \xi_1 + \int_t^T f_s^1(Z_s^1)\mathrm{d}s - \int_t^T Z_s^1 \cdot \mathrm{d}W_s,\ t \in [0,T]. \tag{11.7}$$

Under Condition (11.2) and Assumption 11.3, there is a unique solution (Y^1, Z^1) with bounded Y^1 and $Z^1 \in \mathbb{H}^2_{\mathrm{BMO}}$. This is achieved by applying Theorem 11.6 under a measure $\mathbb{Q}$ defined by the density $\mathcal{E}(\int_0^{\cdot} Df_t(Z_t^0) \cdot \mathrm{d}W_t)$ where $Z^0 := 0$ and $Df_t(0)$ is bounded. See also Lemma 11.8.

Step 2. Given $(Y^j, Z^j)_{j \le i-1}$, we define the generator

$$f_t^i(z) := f_t\left(\overline{Z}_t^{i-1} + z\right) - f_t\left(\overline{Z}_t^{i-1}\right) \text{ where } \overline{Z}_t^{i-1} := \sum_{j=1}^{i-1} Z_t^j. \tag{11.8}$$

We will justify in Lemma 11.8 that there is a unique solution (Y^i, Z^i) to the BSDE

$$Y_t^i = \xi_i + \int_t^T f_s^i(Z_s^i)\mathrm{d}s - \int_t^T Z_s^i \cdot \mathrm{d}W_s,\ t \in [0,T], \tag{11.9}$$

with bounded Y^i and such that $\overline{Z}^i := Z^1 + \cdots + Z^i \in \mathbb{H}^2_{\mathrm{BMO}}$.

Step 3. We finally observe that by setting $Y := Y^1 + \cdots + Y^n$, $Z := \overline{Z}^n$, and by summing the BSDEs (11.9), we directly obtain

$$\begin{aligned} Y_t &= \sum_{i=1}^n \xi_i + \int_t^T \sum_{i=1}^n f_s^i(Z_s^i)\mathrm{d}s - \int_t^T Z_s \cdot \mathrm{d}W_s \\ &= \xi + \int_t^T f_s(Z_s)\mathrm{d}s - \int_t^T Z_s \cdot \mathrm{d}W_s, \end{aligned}$$

which means that (Y, Z) is a solution of our quadratic BSDE of interest. Moreover, Y inherits the boundedness of the Y^i's, and therefore $Z \in \mathbb{H}^2_{\mathrm{BMO}}$ by Lemma 11.2. Finally, as mentioned before, uniqueness is a consequence of Theorem 11.5.

By the above argument, we have the following existence and uniqueness result.

Theorem 11.7. *Let f be a quadratic generator satisfying* (11.2) *and Assumption 11.3. Then, for any $\xi \in \mathbb{L}^\infty(\mathcal{F}_T)$, there is a unique solution $(Y, Z) \in \mathcal{S}^\infty \times \mathbb{H}^2_{\mathrm{BMO}}$ to the quadratic BSDE* (11.1).

For the proof of this theorem, it only remains to show the existence claim in Step 2.

Lemma 11.8. *For $i = 1, \ldots, n$, let the final data ξ^i be bounded as in* (11.6). *Then there exists a unique solution $(Y^i, Z^i)_{1 \le i \le n}$ of the BSDEs* (11.9) *with bounded Y^i's. Moreover, the process $\overline{Z}^i := Z^0 + \cdots + Z^i \in \mathbb{H}^2_{\mathrm{BMO}}$ for all $i = 1, \ldots, n$.*

Proof. We shall argue by induction. That the claim is true for $i = 1$ was justified in Step 1 above by following exactly the same argument as in Step 2. We next assume that the claim is true for all $j \leq i - 1$ and extend it to i.

1. We first prove a convenient estimate for the generator. Set

$$\phi_t^i := Df_t^i(0) = Df_t(\overline{Z}^{i-1}). \tag{11.10}$$

Then, it follows from the mean value theorem that there exists a random $\lambda = \lambda(\omega \in [0,1]$ such that

$$\begin{aligned}\left|f_t^i(z) - f_t^i(z') - \phi_t^i \cdot (z-z')\right| &= \left|Df_t^i(\lambda z + (1-\lambda)z') - Df_t^i(0)\right| \, |z-z'| \\ &\leq \theta_2 |\lambda z + (1-\lambda)z'| \, |z-z'| \\ &\leq \theta_2 |z-z'|(|z| + |z'|). \end{aligned} \tag{11.11}$$

2. We rewrite the BSDE (11.9) into

$$Y_t^i = \xi_i + \int_t^T h_s^i(Z_s^i)\mathrm{d}s - \int_t^T Z_s^i \cdot (\mathrm{d}W_s - \phi_s^i \mathrm{d}s), \text{ where } h_s^i(z) := f_s^i(z) - \phi_s^i \cdot z.$$

By the definition of the process ϕ^i in (11.10), it follows from Assumption 11.3 that $|\phi_t^i| \leq \theta_1(1 + |\overline{Z}_t^{i-1}|)$. Then $\phi^i \in \mathbb{H}^2_{\mathrm{BMO}}$ is inherited from the induction hypothesis which guarantees that $Z^j \in \mathbb{H}^2_{\mathrm{BMO}}$ for $j \leq i - 1$, and therefore $\overline{Z}^{i-1} \in \mathbb{H}^2_{\mathrm{BMO}}$. By Property 3 of BMO martingales, we then conclude that

$$B^i := W - \int_0^{\cdot} \phi_s^i \mathrm{d}s \text{ is a Brownian motion under } \mathbb{Q}^i := \mathcal{E}\left(\int_0^{\cdot} \phi_s^i \cdot \mathrm{d}W_s\right)_T \cdot \mathbb{P}.$$

We now view the latter BSDE as formulated under the equivalent probability measure $\mathbb{Q}^i$ by

$$Y_t^i = \xi_i + \int_t^T h_s^i(Z_s^i)\mathrm{d}s - \int_t^T Z_s^i \cdot \mathrm{d}B_s^i, \; \mathbb{Q}^i - \text{a.s.}$$

where, by (11.11), the quadratic generator h^i satisfies the conditions of Theorem 11.6 with the same parameter θ_2 and the existence of a unique solution $(Y^i, Z^i) \in \mathcal{S}^\infty \times \mathbb{H}^2_{\mathrm{BMO}}(\mathbb{Q}^i)$ follows.

3. It remains to prove that $\overline{Z}^i := Z^1 + \cdots + Z^i \in \mathbb{H}^2_{\mathrm{BMO}}$. To see this, we define $\overline{Y}^i := Y^1 + \cdots + Y^i$, and observe that the pair process $(\overline{Y}^i, \overline{Z}^i)$ solves the BSDE

$$\overline{Y}_t^i = \sum_{j=1}^i \xi_j + \int_t^T \sum_{j=1}^i f_s^j(Z_s^j)\mathrm{d}s - \int_t^T \overline{Z}_s^i \cdot \mathrm{d}W_s$$

$$= \sum_{j=1}^{i} \xi_j + \int_t^T f_s^i(\overline{Z}_s^i)\mathrm{d}s - \int_t^T \overline{Z}_s^i \cdot \mathrm{d}W_s.$$

Since $\sum_{j=1}^{i} \xi_j$ is bounded and f^i satisfies (11.2), it follows from Lemma 11.2 that $\overline{Z}^i \in \mathbb{H}^2_{\mathrm{BMO}}$. □

Remark 11.9. The conditions of Assumption 11.3 can be weakened by essentially removing the smoothness conditions. Indeed an existence result was established by Kobylansky [26] and Morlais [31] under weaker assumptions.

11.3 Portfolio Optimization Under Constraints

The application of this section was first introduced by Elkaroui and Rouge [17] and Imkeller et al. [22].

11.3.1 Problem Formulation

In this section, we consider a financial market consisting of a non-risky asset, normalized to unity, and d risky assets $S = (S^1, \ldots, S^d)$ defined by some initial condition S_0 and the dynamics:

$$\mathrm{d}S_t = S_t \star \sigma_t \left(\mathrm{d}W_t + \theta_t \mathrm{d}t\right),$$

where θ and σ are bounded progressively measurable processes with values in $\mathbb{R}^d$ and $\mathbb{R}^{d\times d}$, respectively. We also assume that σ_t is invertible with bounded inverse process σ^{-1}.

In financial words, θ is the risk premium process, and σ is the volatility (matrix) process.

Given a maturity $T > 0$, a portfolio strategy is a progressively measurable process $\{\pi_t, t \leq T\}$ with values in $\mathbb{R}^d$ and such that $\int_0^T |\pi_t|^2 \mathrm{d}t < \infty$, $\mathbb{P}$–a.s.

For each $i = 1, \ldots, d$ and $t \in [0, T]$, π_t^i denotes the dollar amount invested in the ith risky asset at time t. Then, the liquidation value of a self-financing portfolio defined by the portfolio strategy π and the initial capital X_0 is given by

$$X_t^\pi = X_0 + \int_0^t \pi_r \cdot \sigma_r \left(\mathrm{d}W_r + \theta_r \mathrm{d}r\right), \; t \in [0, T]. \tag{11.12}$$

We shall impose more conditions later on the set of portfolio strategies. In particular, we will consider the case where the portfolio strategy is restricted to some

$$A \text{ closed convex subset of } \mathbb{R}^d.$$

The objective of the portfolio manager is to maximize the expected utility of the final liquidation value of the portfolio, where the utility function is defined by

$$U(x) := -\mathrm{e}^{-x/\eta} \quad \text{for all } x \in \mathbb{R}, \tag{11.13}$$

for some parameter $\eta > 0$ representing the risk tolerance of the investor, i.e., η^{-1} is the risk aversion.

Definition 11.10. A portfolio strategy $\pi \in \mathbb{H}^2_{\mathrm{loc}}$ is said to be admissible if it takes values in A and

$$\text{the family } \left\{\mathrm{e}^{-X^\pi_\tau/\eta},\ \tau \in \mathcal{T}_0^T\right\} \text{ is uniformly integrable.} \tag{11.14}$$

We denote by $\mathcal{A}$ the collection of all admissible portfolio strategies.

We are now ready for the formulation of the portfolio manager problem. Let ξ be some bounded $\mathcal{F}_T$−measurable r.v. representing the liability at the maturity T. The portfolio manager problem is defined by the stochastic control problem:

$$V_0 := \sup_{\pi \in \mathcal{A}} \mathbb{E}\left[U\left(X^\pi_T - \xi\right)\right]. \tag{11.15}$$

Our main objective in the subsequent subsections is to provide a characterization of the value function and the solution of this problem in terms of a BSDE.

Remark 11.11. The restriction to the exponential utility case (11.13) is crucial to obtain a connection of this problem to BSDEs.

- In the Markovian framework, we may characterize the value function V by means of the corresponding dynamic programming equation. Then, extending the definition in a natural way to allow for a changing time origin, the dynamic programming equation of this problem is

$$-\partial_t v - \sup_\pi \left\{\pi \cdot \sigma\theta D_x v + \frac{1}{2}|\sigma^{\mathrm{T}}\pi|^2 D_{xx} v + (\sigma^{\mathrm{T}}\pi)\cdot(s \star \sigma D_{xs} v)\right\} = 0. \tag{11.16}$$

  Notice that the above PDE is fully nonlinear, while BSDEs are connected to semilinear PDEs. So, in general, there is no reason for the portfolio optimization problem to be related to BSDEs.
- Let us continue the discussion of the Markovian framework in the context of an exponential utility. Due to the expression of the liquidation value process (11.13), it follows that $U(X_T^{X_0,\pi}) = \mathrm{e}^{-X_0/\eta} U(X_T^{0,\pi})$, where we emphasized the dependence of the liquidation value on the initial capital X_0. Then, by definition of the value function V, we have

$$V(t,x,s) = \mathrm{e}^{-x/\eta} V(t,0,s),$$

i.e., the dependence of the value function V in the variable x is perfectly determined. By plugging this information into the dynamic programming equation (11.16), it turns out that the resulting PDE for the function $U(t,s) := V(t,0,s)$ is semilinear, thus explaining the connection to BSDEs.

- A similar argument holds true in the case of power utility function $U(x) = x^p/p$ for $p < 1$. In this case, due to the domain restriction of this utility function, one defines the wealth process X in a multiplicative way, by taking as control $\tilde{\pi}_t := \pi_t / X_t$, the proportion of wealth invested in the risky assets. Then, it follows that $X_T^{X_0,\tilde{\pi}} = X_0 X_T^{1,\tilde{\pi}}$, $V(t,x,s) = x^p V(t,0,s)$ and the PDE satisfied by $V(t,0,s)$ turns out to be semilinear.

11.3.2 BSDE Characterization

The main result of this section provides a characterization of the portfolio manager problem in terms of the BSDE:

$$Y_t = \xi + \int_t^T f_r(Z_r)\mathrm{d}r - \int_t^T Z_r \cdot \mathrm{d}W_r,\ t \leq T, \tag{11.17}$$

where the generator f is given by

$$\begin{aligned} f_t(z) &:= -z \cdot \theta_t - \frac{\eta}{2}|\theta_t|^2 + \frac{1}{2\eta} \inf_{\pi \in A} \left|\sigma_t^{\mathrm{T}} \pi - (z + \eta\theta_t)\right|^2. \\ &= -z \cdot \theta_t - \frac{\eta}{2}|\theta_t|^2 + \frac{1}{2\eta} \mathrm{dist}(z + \eta\theta_t, \sigma_t A)^2, \end{aligned} \tag{11.18}$$

where for $x \in \mathbb{R}^d$, $\mathrm{dist}(x, \sigma_t A)$ denotes the Euclidean distance from x to the set $\sigma_t A$, the image of A by the matrix σ_t.

Example 11.12 (Complete market). Consider the case $A = \mathbb{R}^d$, i.e., no portfolio constraints. Then $f_t(z) = -z \cdot \theta_t - \frac{\eta}{2}|\theta_t|^2$ is an affine generator in z, and the above BSDE can be solved explicitly:

$$Y_t = \mathbb{E}_t^{\mathbb{Q}}\left[\xi - \frac{\eta}{2}\int_t^T |\theta_r|^2 \mathrm{d}r\right],\ t \in [0,T],$$

where $\mathbb{Q}$ is the so-called risk-neutral probability measure which turns the process $W + \int_0^{\cdot} \theta_r \mathrm{d}r$ into a Brownian motion.

Notice that, except for the complete market case $A = \mathbb{R}^d$ of the previous example, the above generator is always quadratic in z. See however Exercise 11.15 for another explicitly solvable example.

Since the risk premium process is assumed to be bounded, the above generator satisfies Condition (11.2). As for Assumption 11.3, its verification depends on the geometry of the set A. Finally, the final condition represented by the liability ξ is assumed to be bounded.

Theorem 11.13. *Let A be a closed convex set, and suppose that f satisfies Assumption 11.3. Then the value function of the portfolio management problem and the corresponding optimal portfolio are given by*

$$V_0 = -e^{-\frac{1}{\eta}(X_0 - Y_0)} \text{ and } \hat{\pi}_t := \text{Arg}\min_{\pi \in A} |\sigma_t^{\mathrm{T}}\pi - (Z_t + \eta\theta_t)|,$$

where X_0 is the initial capital of the investor and (Y, Z) is the unique solution of the quadratic BSDE (11.17).

Proof. For every $\pi \in \mathcal{A}$, we define the process

$$V_t^\pi := -e^{-(X_t^{0,\pi} - Y_t)/\eta}, \ t \in [0, T].$$

1. We first compute by Itô's formula that

$$\begin{aligned} dV_t^\pi &= -\frac{1}{\eta} V_t^\pi \left(dX_t^{0,\pi} - dY_t \right) + \frac{1}{2\eta^2} V_t^\pi d\langle X^{0,\pi} - Y \rangle_t \\ &= -\frac{1}{\eta} V_t^\pi \Big[(f_t(Z_t) - \varphi_t(Z_t, \pi_t))\, dt + \left(\sigma_t^{\mathrm{T}} \pi_t - Z_t \right) \cdot dW_t \Big], \end{aligned}$$

where we denoted:

$$\begin{aligned} \varphi_t(z, \pi) &:= -\sigma_t^{\mathrm{T}} \pi \cdot \theta_t + \frac{1}{2\eta} |\sigma_t^{\mathrm{T}} \pi_t - z|^2 \\ &= -z \cdot \theta_t - \frac{\eta}{2} |\theta_t|^2 + \frac{1}{2\eta} \left| \sigma_t^{\mathrm{T}} \pi - (z + \eta\theta_t) \right|^2, \end{aligned}$$

so that $f_t(z) = \inf_{\pi \in A} \varphi_t(z, \pi)$. Consequently, the process V^π is a local supermartingale. Now recall from Theorem 11.7 that the solution (Y, Z) of the quadratic BSDE has a bounded component Y. Then, it follows from admissibility condition (11.14) of Definition 11.10 that the process V^π is a supermartingale. In particular, this implies that $-e^{-(X_0 - Y_0)/\eta} \geq \mathbb{E}[V_T^\pi]$, and it then follows from the arbitrariness of $\pi \in \mathcal{A}$ that

$$V_0 \leq -e^{-(X_0 - Y_0)/\eta}. \tag{11.19}$$

2. To prove the reverse inequality, we notice that the portfolio strategy $\hat{\pi}$ introduced in the statement of the theorem satisfies

$$dV_t^{\hat{\pi}} = -\frac{1}{\eta} V_t^{\hat{\pi}} \left(\sigma_t^{\mathrm{T}} \hat{\pi}_t - Z_t \right) \cdot dW_t.$$

Then $V_t^{\hat{\pi}}$ is a local martingale. We continue by estimating its diffusion part:

$$\begin{aligned}\left|\sigma_t^{\mathrm{T}}\hat{\pi}_t - Z_t\right| &\leq \eta|\theta_t| + \left|\sigma_t^{\mathrm{T}}\hat{\pi}_t - (Z_t + \eta\theta_t)\right| \\ &= \eta|\theta_t| + \sqrt{f_t(Z_t) + Z_t\cdot\theta_t + \frac{\eta}{2}|\theta_t|^2} \\ &\leq C(1+|Z_t|),\end{aligned}$$

for some constant C. Since $Z \in \mathbb{H}^2_{\mathrm{BMO}}$ by Theorem 11.7, this implies that $\sigma_t^{\mathrm{T}}\hat{\pi}_t - Z_t \in \mathbb{H}^2_{\mathrm{BMO}}$ and $\sigma_t^{\mathrm{T}}\hat{\pi}_t \in \mathbb{H}^2_{\mathrm{BMO}}$. Then, it follows from Property 3 of BMO martingales that $\hat{\pi} \in \mathcal{A}$ and $V^{\hat{\pi}}$ is a martingale. Hence,

$$\hat{\pi} \in \mathcal{A} \text{ and } \mathbb{E}\left[-\mathrm{e}^{-(X_T^{\hat{\pi}} - Y_T)/\eta}\right] = -\mathrm{e}^{-(X_0 - Y_0)/\eta}$$

which, together with (11.19), shows that $V_0 = -\mathrm{e}^{-(X_0-Y_0)/\eta}$ and $\hat{\pi}$ is an optimal portfolio strategy. □

Remark 11.14. The condition that A is convex in Theorem 11.13 can be dropped by defining the optimal portfolio process $\hat{\pi}$ as a measurable selection in the set of minimizers of the norm $|\sigma_t^{\mathrm{T}}\pi - (Z_t + \eta\theta_t)|$ over $\pi \in A$. See Imkeller et al. [22].

Exercise 11.15. The objective of the following problem is to provide an example of portfolio optimization problem in incomplete market which can be explicitly solved. This is a non-Markovian version of the PDE based work of Zariphopoulou [42].

1. *Portfolio optimization under stochastic volatility* Let $W = (W^1, W^2)$ be a standard Brownian motion on a complete probability space $(\Omega, \mathcal{F}, \mathbb{P})$, and denote $\mathbb{F}^i := \{\mathcal{F}_t^i = \sigma(W_s^i, s \leq t)\}_{t\geq 0}$, $\mathbb{F} := \{\mathcal{F}_t = \mathcal{F}_t^1 \vee \mathcal{F}_t^2\}_{t\geq 0}$.
   Consider the portfolio optimization problem:

$$V_0 := \sup_{\pi\in\mathcal{A}} \mathbb{E}\Big[-\mathrm{e}^{-\eta(X_T^\pi - \xi)}\Big],$$

where $\eta > 0$ is the absolute risk-aversion coefficient, ξ is a bounded $\mathcal{F}_T$−measurable random variable, and

$$X_T^\pi := \int_0^T \pi_t\sigma_t\Big(\rho_t \mathrm{d}W_t^1 + \sqrt{1-\rho_t^2}\mathrm{d}W_t^2 + \theta_t \mathrm{d}t\Big)$$

is the liquidation value at time T of a self-financing portfolio π in the financial market with zero interest rates and stock price defined by the risk premium and the volatility processes θ and σ. The latter processes are $\mathbb{F}$−bounded and progressively measurable. Finally, ρ is a correlation process taking values in $[0, 1]$, and the admissibility set $\mathcal{A}$ is defined as in Imkeller-Hu-Müller (see my Fields Lecture notes).

2. *BSDE characterization.* This problem fits in the framework of Hu-Imkeller-Müller of portfolio optimization under constrained portfolio (here the portfolio is constrained to the closed convex subset $\mathbb{R} \times \{0\}$ of $\mathbb{R}^2$).

   We introduce a risk-neutral measure $\mathbb{Q}$ under which the process $B = (B^1, B^2)$:

$$B_t^1 := W_t^1 + \int_0^t \theta_r \rho_r \mathrm{d}r \text{ and } B_t^2 := W_t^2 + \int_0^t \theta_r \sqrt{1-\rho_t^2} \mathrm{d}r,$$

   is a Brownian motion.

   Then, it follows that $V_0 = \mathrm{e}^{\eta Y_0}$, where (Y, Z) is the unique solution of the quadratic BSDE:

$$Y_t = \xi + \int_t^T f_r(Z_r)\mathrm{d}r - \int_t^T Z_r \cdot \mathrm{d}B_r \qquad (11.20)$$

   where the generator $f : \mathbb{R}_+ \times \Omega \times \mathbb{R}^2$ is defined by

$$f_t(z) := -\frac{\theta_t^2}{2\eta} + \frac{\eta}{2}\left(\sqrt{1-\rho_t^2}\, z_1 + \rho_t z_2\right)^2 \quad \text{for all } z \in \mathbb{R}^2.$$

   The existence of a unique solution to this BSDE with bounded component Y and BMO martingale $\int_0^\cdot Z_t \cdot \mathrm{d}B_t$ is guaranteed by our results in the present section.

3. *Conditionally Gaussian stochastic volatility.* We next specialize the discussion to the case:

$$\xi \text{ is } \mathcal{F}_T^1\text{-measurable, and } \theta, \sigma, \rho \text{ are } \mathbb{F}^1\text{-progressively measurable.}$$

   Then, by adaptability considerations, it follows that the second component of the Z-process $Z^2 \equiv 0$. Denoting the first component by $\zeta := Z^1$, this reduces the BSDE (11.20) to

$$Y_t = \xi + \int_t^T \left(-\frac{\theta_t^2}{2\eta} + \frac{\eta}{2}(1-\rho_t^2)\zeta_t^2\right)\mathrm{d}r - \int_t^T \zeta_r \mathrm{d}B_r^1. \qquad (11.21)$$

4. *Linearizing the BSDE.* To achieve additional simplification, we further assume that the correlation process ρ is constant. Then for a constant $\beta \in \mathbb{R}$, we immediately compute by Itô's formula that the process $y_t := \mathrm{e}^{\beta Y_t}$ satisfies

$$\frac{\mathrm{d}y_t}{y_t} = \beta\left(\frac{\theta_t^2}{2\eta} - \frac{\eta}{2}(1-\rho^2)\zeta_t^2\right)\mathrm{d}t + \frac{1}{2}\beta^2\zeta_t^2\mathrm{d}t + \beta\zeta_t \mathrm{d}B_t$$

   so that the choice

$$\beta^2 := \eta(1-\rho^2)$$

leads to a constant generator for y. We now continue in the obvious way representing y_0 as an expected value and deducing Y_0.

5. *Utility indifference.* In the present framework, we may compute explicitly the utility indifference price of the claim ξ.... This leads to a nonlinear pricing rule which has nice financial interpretations.

11.4 Interacting Investors with Performance Concern

11.4.1 The Nash Equilibrium Problem

In this section, we consider N portfolio managers $i = 1, \ldots, N$ whose preferences are characterized by expected exponential utility functions with tolerance parameters η_i:

$$U^i(x) := -\mathrm{e}^{-x/\eta^i}, \; x \in \mathbb{R}. \tag{11.22}$$

In addition, we assume that each investor is concerned about the average performance of his peers. Given the portfolio strategies $\pi^i, i = 1, \ldots, N$, of the managers, we introduce the average performance viewed by agent i as

$$\overline{X}^{i,\pi} := \frac{1}{N-1} \sum_{j \neq i} X_T^{\pi^j}. \tag{11.23}$$

The portfolio optimization problem of the ith agent is then defined by

$$V_0^i\left((\pi^j)_{j\neq i}\right) := V_0^i$$
$$:= \sup_{\pi^i \in \mathcal{A}^i} \mathbb{E}\left[U^i\left((1-\lambda^i)X_T^{\pi^i} + \lambda^i(X_T^{\pi^i} - \overline{X}_T^{i,\pi})\right)\right], \; 1 \leq i \leq N, \tag{11.24}$$

where $\lambda^i \in [0,1]$ measures the sensitivity of agent i to the performance of his peers, and the set of admissible portfolios $\mathcal{A}^i$ is defined as follows.

Definition 11.16. A progressively measurable process π^i with values in $\mathbb{R}^d$ is said to be admissible for agent i, and we denote $\pi^i \in \mathcal{A}^i$ if:

1. π^i takes values in A_i, a given closed convex subset of $\mathbb{R}^d$.
2. $\mathbb{E}[\int_0^T |\pi_t^i|^2 \mathrm{d}t] < \infty$,
3. The family $\left\{\mathrm{e}^{-X_\tau^{\pi^i}/\eta^i}, \tau \in \mathcal{T}\right\}$ is uniformly bounded in $\mathbb{L}^p$ for all $p > 1$.

Our main interest is to find a Nash equilibrium, i.e., a situation where all portfolio managers are happy with the portfolio given those of their peers.

Definition 11.17. A Nash equilibrium for the N portfolio managers is an N-tuple $(\hat{\pi}^1, \ldots, \hat{\pi}^N) \in \mathcal{A}^1 \times \cdots \times \mathcal{A}^N$ such that, for every $i = 1, \ldots, N$, given $(\hat{\pi}^j)_{j \neq i}$, the portfolio strategy $\hat{\pi}^i$ is a solution of the portfolio optimization problem $V_0^i\left((\hat{\pi}^j)_{j \neq i}\right)$.

11.4.2 The Individual Optimization Problem

In this section, we provide a formal argument which helps to understand the construction of Nash equilibrium of the subsequent section.

For fixed $i = 1, \ldots, N$, we rewrite (11.24) as

$$V_0^i := \sup_{\pi^i \in \mathcal{A}^i} \mathbb{E}\left[U^i\left(X_T^{\pi^i} - \tilde{\xi}^i\right)\right], \text{ where } \tilde{\xi}^i := \lambda^i \overline{X}_T^{i,\pi}. \tag{11.25}$$

Then, from the example of the previous section, we expect that value function V_0^i and the corresponding optimal solution be given by

$$V_0^i = -\mathrm{e}^{-(X_0^i - \tilde{Y}_0^i)/\eta^i}, \tag{11.26}$$

and

$$\sigma_t^{\mathrm{T}} \hat{\pi}_t^i = \mathbf{a}_t^i(\tilde{\zeta}_t^i + \eta^i \theta_t) \text{ where } \mathbf{a}_t^i(z^i) := \operatorname{Arg}\min_{u^i \in A^i} |\sigma_t^{\mathrm{T}} u^i - z^i|, \tag{11.27}$$

and $(\tilde{Y}^i, \tilde{\zeta}^i)$ is the solution of the quadratic BSDE:

$$\tilde{Y}_t^i = \tilde{\xi}^i + \int_t^T \left(-\tilde{\zeta}_r^i \cdot \theta_r - \frac{\eta^i}{2}|\theta_r|^2 + \tilde{f}_r^i(\tilde{\zeta}_r^i + \eta^i \theta_r)\right)\mathrm{d}r - \int_t^T \tilde{\zeta}_r^i \cdot \mathrm{d}W_r,\ t \le T, \tag{11.28}$$

and the generator $\tilde{f}^i$ is given by

$$\tilde{f}_t^i(z^i) := \frac{1}{2\eta^i}\operatorname{dist}(z^i, \sigma_t A^i)^2,\ z^i \in \mathbb{R}^d. \tag{11.29}$$

This suggests that one can search for a Nash equilibrium by solving the BSDEs (11.28) for all $i = 1, \ldots, N$. However, this raises the following difficulties.

The first concern that one would have is that the final data $\tilde{\xi}^i$ does not have to be bounded as it is defined in (11.25) through the performance of the other portfolio managers.

But in fact, the situation is even worse because the final data $\tilde{\xi}^i$ induces a coupling of the BSDEs (11.28) for $i = 1, \ldots, N$. To express this coupling in a more transparent way, we substitute the expressions of $\tilde{\xi}^i$ and rewrite (11.28) for $t = 0$ into:

$$\tilde{Y}_0^i = \eta^i \xi + \int_0^T \tilde{f}_r^i(\zeta_r^i)\mathrm{d}r - \int_0^T \left(\zeta_r^i - \lambda_N^i \sum_{j \neq i} \mathbf{a}_r^j(\zeta_r^j) \right) \cdot \mathrm{d}B_r$$

where the process $B := W + \int_0^\cdot \theta_r \mathrm{d}r$ is the Brownian motion under the equivalent martingale measure,

$$\lambda_N^i := \frac{\lambda^i}{N-1}, \; \zeta_t^i := \tilde{\zeta}_t^i + \eta^i \theta_t, \; t \in [0,T],$$

and the final data is expressed in terms of the unbounded r.v.

$$\xi := \int_0^T \theta_r \cdot \mathrm{d}B_r - \frac{1}{2}\int_0^T |\theta_t|^2 \mathrm{d}t.$$

Then $\tilde{Y}_0 = Y_0$, where (Y, ζ) is defined by the BSDE

$$Y_t^i = \eta^i \xi + \int_t^T \tilde{f}_r^i(\zeta_r^i)\mathrm{d}r - \int_t^T \left(\zeta_r^i - \lambda_N^i \sum_{j \neq i} \mathbf{a}_r^j(\zeta_r^j) \right) \cdot \mathrm{d}B_r. \qquad (11.30)$$

In order to sketch (11.30) into the BSDEs framework, we further introduce the mapping $\phi_t : \mathbb{R}^{Nd} \longrightarrow \mathbb{R}^{Nd}$ defined by the components:

$$\phi_t^i(\zeta^1, \ldots, \zeta^N) := \zeta^i - \lambda_N^i \sum_{j \neq i} \mathbf{a}_t^j(\zeta^j) \quad \text{for all } \zeta^1, \ldots, \zeta^N \in \mathbb{R}^d. \qquad (11.31)$$

It turns out that the mapping ϕ_t is invertible under fairly general conditions. We shall prove this result in Lemma 11.18 in the case where the A^i's are linear subspaces of $\mathbb{R}^d$. Then one can rewrite (11.30) as

$$Y_t^i = \eta^i \xi + \int_t^T f_r^i(Z_r)\mathrm{d}r - \int_t^T Z_r^i \cdot \mathrm{d}B_r, \qquad (11.32)$$

where the generator f^i is now given by

$$f^i(z) := \tilde{f}_r^i\left(\{\phi_t^{-1}(z)\}^i\right) \quad \text{for all } z = (z^1, \ldots, z^N) \in \mathbb{R}^{Nd}, \qquad (11.33)$$

and $\{\phi_t^{-1}(z)\}^i$ indicates the ith block component of size d of $\phi_t^{-1}(z)$.

11.4.3 The Case of Linear Constraints

We now focus on the case where the constraints sets are such that

$$A^i \text{ is a linear subspace of } \mathbb{R}^d, \; i = 1, \ldots, N. \qquad (11.34)$$

Then, denoting by P_t^i the orthogonal projection operator on $\sigma_t A^i$ (i.e., the image of A^i by the matrix σ_t), we immediately compute that

$$\mathbf{a}_t^i(\zeta^i) := P_t^i(\zeta^i) \tag{11.35}$$

and

$$\phi_t^i(\zeta^1, \ldots, \zeta^N) := \zeta^i - \lambda_N^i \sum_{j \neq i} P_t^j(\zeta^j), \quad \text{for } i = 1, \ldots, N. \tag{11.36}$$

Lemma 11.18. *Let $(A^i)_{1 \le i \le N}$ be linear subspaces of $\mathbb{R}^d$. Then, for all $t \in [0, T]$:*

(i) *The linear mapping ϕ_t of (11.36) is invertible if and only if*

$$\prod_{i=1}^N \lambda^i < 1 \text{ or } \bigcap_{i=1}^N A^i = \{0\}. \tag{11.37}$$

(ii) *This condition is equivalent to the invertibility of the matrices $I_d - Q_t^i$, $i = 1, \ldots, N$, where*

$$Q_t^i := \sum_{j \neq i} \frac{\lambda_N^j}{1 + \lambda_N^j} P_t^j (I_d + \lambda_N^j P_t^i).$$

(iii) *Under (11.37), the ith component of ϕ_t^{-1} is given by*

$$\{\phi_t^{-1}(z)\}^i = (I_d - Q_t^i)^{-1} \left(z^i + \sum_{j \neq i} \frac{1}{1 + \lambda_N^j} P_t^j (\lambda_N^i z^j - \lambda_N^j z^i) \right).$$

Proof. We omit all t subscripts, and we denote $\mu^i := \lambda_N^i$. For arbitrary $z^1, \ldots, z^N$ in $\mathbb{R}^d$, we want to find a unique solution to the system

$$\zeta^i - \mu^i \sum_{j \neq i} P^j \zeta^j = z^i, \; 1 \le i \le N. \tag{11.38}$$

1. Since P^j is a projection, we immediately compute that $(I_d + \mu^j P^j)^{-1} = I_d - \frac{\mu^j}{1+\mu^j} P^j$. Subtracting equations i and j from the above system, we see that

$$\begin{aligned} \mu^i P^j \zeta^j &= P^j (I_d + \mu^j P^j)^{-1} \Big(\mu^j (I_d + \mu^i P^i) \zeta^i + \mu^i zj - \mu^j z^i \Big) \\ &= \frac{1}{1 + \mu^j} P^j \Big(\mu^j (I_d + \mu^i P^i) \zeta^i + \mu^i z^j - \mu^j z^i \Big). \end{aligned}$$

Then it follows from (11.38) that

$$z^i = \zeta^i - \sum_{j \neq i} \frac{1}{1+\mu^j} P^j \left(\mu^j (I_d + \mu^i P^i)\zeta^i + \mu^i z^j - \mu^j z^i \right),$$

and we can rewrite (11.38) equivalently as

$$\left(I_d - \sum_{j \neq i} \frac{\mu^j}{1+\mu^j} P^j (I_d + \mu^i P^i) \right) \zeta^i = z^i + \sum_{j \neq i} \frac{1}{1+\mu^j} P^j (\mu^i z^j - \mu^j z^i), \tag{11.39}$$

so that the invertibility of ϕ is now equivalent to the invertibility of the matrices $I_d - Q^i, i = 1, \ldots, N$, where Q^i is introduced in statement of the lemma.

2. We now prove that $I_d - Q^i$ is invertible for every $i = 1, \ldots, N$ iff (11.37) holds true.

   (a) First, assume to the contrary that $\lambda^i = 1$ for all i and $\cap_{i=1}^N A^i$ contains a nonzero element x^0. Then, it follows that $y^0 := \sigma^{\mathrm{T}} x^0$ satisfies $P^i y^0 = y^0$ for all $i = 1, \ldots, N$, and therefore $Q^i y^0 = y^0$. Hence, $I_d - Q^i$ is not invertible.

   (b) Conversely, we consider separately two cases.

   - If $\lambda^{i_0} < 1$ for some $i_0 \in \{1, \ldots, N\}$, we estimate that

     $$\frac{\mu^{i_0}}{1+\mu^{i_0}} < \frac{\frac{1}{N-1}}{1+\frac{1}{N-1}} \text{ and } \frac{\mu^i}{1+\mu^i} \leq \frac{\frac{1}{N-1}}{1+\frac{1}{N-1}} \text{ for } i \neq i_0.$$

     Then for all $i \neq i_0$ and $x \neq 0$, it follows that $|Q^i x| < |x|$ proving that $I - Q^i$ is invertible.

   - If $\lambda^i = 1$ for all $i = 1, \ldots, N$, then for all $x \in \mathrm{Ker}(Q^i)$, we have $x = Q^i x$ and therefore

     $$\begin{aligned} |x| &= \left| \sum_{j \neq i} \frac{\mu^j}{1+\mu^j} P^j (I_d + \mu^i P^i) x \right| \\ &= \frac{1}{N} \left| \sum_{j \neq i} P^j (I_d + \frac{1}{N-1} P^i) x \right| \\ &\leq \frac{1}{N} \sum_{j \neq i} (1 + \frac{1}{N-1} |x| = |x|, \end{aligned}$$

     where we used the fact that the spectrum of the P^i's is reduced to $\{0, 1\}$. Then equality holds in the above inequalities, which can only happen if

$P^i x = x$ for all $i = 1, \ldots, N$. We can then conclude that $\cap_{i=1}^N \mathrm{Ker}(I_d - P^i) = \{0\}$ implies that $I_d - Q^i$ is invertible. This completes the proof as $\cap_{i=1}^N \mathrm{Ker}(I_d - P^i) = \{0\}$ is equivalent to $\cap_{i=1}^N A^i = \{0\}$. □

11.4.4 Nash Equilibrium Under Deterministic Coefficients

The discussion of Sect. 11.4.2 shows that the question of finding a Nash equilibrium for our problem reduces to the vector BSDE with quadratic generator (11.32), that we rewrite here for convenience:

$$Y_t^i = \eta^i \xi + \int_t^T f_r^i(Z_r)\mathrm{d}r - \int_t^T Z_r^i \cdot \mathrm{d}B_r, \tag{11.40}$$

where $\xi := \int_0^T \theta_r \cdot \mathrm{d}B_r - \frac{1}{2}\int_0^T |\theta_r|^2 \mathrm{d}r$, and the generator f^i is given by:

$$f^i(z) := \tilde{f}_r^i\left(\{\phi_t^{-1}(z)\}^i\right) \quad \text{for all } z = (z^1, \ldots, z^N) \in \mathbb{R}^{Nd}. \tag{11.41}$$

Unfortunately, the problem of solving vector BSDEs with quadratic generator is still not understood. Therefore, we will not continue in the generality assumed so far, and we will focus in the sequel on the case where

$$\text{the } A^i\text{'s are vector subspaces of } \mathbb{R}^d \text{ and}$$
$$\sigma_t = \sigma(t) \text{ and } \theta_t = \theta(t) \text{ are deterministic functions.} \tag{11.42}$$

Then, the vector BSDE reduces to

$$Y_t^i = \eta^i \xi + \frac{1}{2\eta^i}\int_t^T \left|(I_d - P^i(t))\left(\{\phi(t)^{-1}(Z_r)\}^i\right)\right|^2 \mathrm{d}r - \int_t^T Z_r^i \cdot \mathrm{d}B_r, \tag{11.43}$$

where $P_t^i = P^i(t)$ is deterministic and $\{\phi_t^{-1}(z)\}^i = \{\phi(t)^{-1}(z)\}^i$ is deterministic and given explicitly by Lemma 11.18(iii).

In this case, an explicit solution of the vector BSDE is given by

$$Z_t^i = \eta^i \theta(t)$$

$$Y_t^i = -\frac{\eta^i}{2}\int_0^T |\theta(t)|^2 \mathrm{d}t + \frac{1}{2\eta^i}\int_0^t |(I_d - P^i(t))M^i(t)\theta(t)|^2 \mathrm{d}t, \tag{11.44}$$

where

$$M^i(t) := \left(I_d - \sum_{j \neq i} \frac{\lambda_N^j}{1+\lambda_N^j} P^j(t)(I_d + \lambda_N^j P^i(t) \right)^{-1}$$
$$\times \left(\eta^i I_d + \sum_{j \neq i} \frac{1}{1+\lambda_N^j} P^j(t)(\lambda_N^i \eta^j - \lambda_N^j \eta^i) \right).$$

By (11.27), the candidate for agent ith optimal portfolio is also deterministic and given by

$$\hat{\pi}^i := \sigma^{-1} P^i M^i \theta, \; i = 1, \ldots, N. \tag{11.45}$$

Proposition 11.19. *In the context of the financial market with deterministic coefficients* (11.42)*, the* N*−tuple* $(\hat{\pi}^1, \ldots, \hat{\pi}^N)$ *defined by* (11.45) *is a Nash equilibrium.*

Proof. The above explicit solution of the vector BSDE induces an explicit solution $(\tilde{Y}^i, \tilde{\zeta}^i)$ of the coupled system of BSDEs (11.28), $1 \le i \le N$ with deterministic $\tilde{\zeta}^i$. In order to prove the required result, we have to argue by verification following the lines of the proof of Theorem 11.13 for every fixed i in $\{1, \ldots, n\}$.

1. First for an arbitrary π^i, we define the process

$$V_t^{\pi^i} := -\mathrm{e}^{-(X_t^{\pi^i} - \tilde{Y}_t^i)/\eta^i}, \; t \in [0, T].$$

By Itô's formula, it is immediately seen that this process is a local supermartingale (the generator has been defined precisely to satisfy this property!). By the admissibility condition of Definition 11.16 together with the fact that $\tilde{Y}^i$ has a Gaussian distribution (as a diffusion process with deterministic coefficients), it follows that the family $\{V_\tau^{\pi^i}, \tau \in \mathcal{T}\}$ is uniformly bounded in $\mathbb{L}^{1+\varepsilon}$ for any $\varepsilon > 0$. Then the process V^{π^i} is a supermartingale. By the arbitrariness of $\pi^i \in \mathcal{A}^i$, this provides the first inequality

$$-\mathrm{e}^{-(X_0^i - \tilde{Y}_0^i)/\eta^i} \ge V_0^i\left((\hat{\pi}^j)_{j \neq i}\right).$$

2. We next prove that equality holds by verifying that $\hat{\pi}^i \in \mathcal{A}^i$ and the process $V^{\hat{\pi}^i}$ is a martingale. This will provide the value function of agent i's portfolio optimization problem and the fact that $\hat{\pi}^i$ is optimal for the problem $V_0^i\left((\hat{\pi}^j)_{j \neq i}\right)$.

That $\hat{\pi}^i \in \mathcal{A}^i$ is immediate; recall again that $\hat{\pi}^i$ is deterministic. As in the previous step, direct application of Itô's formula shows that $V^{\hat{\pi}^i}$ is a local martingale, and the martingale property follows from the fact that $X^{\hat{\pi}^i}$ and $\tilde{Y}^i$ have deterministic coefficients. □

We conclude this section with a simple example which shows the effect of the interaction between managers.

Example 11.20 ($N = 3$ investors, $d = 3$ assets). Consider a financial market with $N = d = 3$. Denoting by (e_1, e_2, e_3) the canonical basis of $\mathbb{R}^3$, the constraints set for the agents are

$$A_1 = \mathbb{R}e_1 + \mathbb{R}e_2,\ A_2 = \mathbb{R}e_2 + \mathbb{R}e_3,\ A_3 = \mathbb{R}e_3,$$

i.e., agent 1 is allowed to trade without constraints the first two assets, agent 2 is allowed to trade without constraints the last two assets, and agent 3 is only allowed to trade the third assets without constraints.

We take $\sigma = I_3$. In the present context of deterministic coefficients, this means that the price processes of the assets are independent. Therefore, if there were no interaction between the investors, their optimal investment strategies would not be affected by the assets that they are not allowed to trade.

In this simple example, all calculations can be performed explicitly. The Nash equilibrium of Proposition 11.19 is given by

$$\begin{aligned}
\hat{\pi}_t^1 &= \eta\theta^1(t)e_1 + \frac{2+\lambda_1}{2-\frac{\lambda_1\lambda_2}{2}}\eta\theta^2(t)e_2,\\
\hat{\pi}_t^2 &= \frac{2+\lambda_2}{2-\frac{\lambda_1\lambda_2}{2}}\eta\theta^2(t)e_2 + \frac{2+\lambda_2}{2-\frac{\lambda_2\lambda_3}{2}}\eta\theta^3(t)e_3, \text{ and}\\
\hat{\pi}_t^3 &= \frac{2+\lambda_3}{2-\frac{\lambda_2\lambda_3}{2}}\eta\theta^3(t)e_3.
\end{aligned}$$

This shows that, whenever two investors have access to the same asset, their interaction induces an overinvestment in this asset characterized by a dilation factor related to the their sensitivity to the performance of the other investor.

Chapter 12
Probabilistic Numerical Methods for Nonlinear PDEs

In this chapter, we introduce a backward probabilistic scheme for the numerical approximation of the solution of a nonlinear partial differential equation. The scheme is decomposed into two steps:

1. The Monte Carlo step consists in isolating the linear generator of some underlying diffusion process, so as to split the PDE into this linear part and a remaining nonlinear one.
2. Evaluating the PDE along the underlying diffusion process, we obtain a natural discrete-time approximation by using finite differences approximation in the remaining nonlinear part of the equation.

Our main concern will be to prove the convergence of this discrete-time approximation. In particular, the above scheme involves the calculation of conditional expectations that should be replaced by some approximation for any practical implementation. The error analysis of this approximation will not be addressed here.

Throughout this chapter, μ and σ are two functions from $\mathbb{R}_+ \times \mathbb{R}^d$ to $\mathbb{R}^d$ and $\mathcal{S}_d$, respectively. Let $a := \sigma^2$, and define the linear operator:

$$\mathcal{L}^X \varphi := \frac{\partial \varphi}{\partial t} + \mu \cdot D\varphi + \frac{1}{2} a \cdot D^2 \varphi.$$

Consider the map

$$F : (t, x, r, p, \gamma) \in \mathbb{R}_+ \times \mathbb{R}^d \times \mathbb{R} \times \mathbb{R}^d \times \mathcal{S}_d \longmapsto F(x, r, p, \gamma) \in \mathbb{R},$$

which is assumed to be elliptic:

$$F(t, x, r, p, \gamma) \geq F(t, x, r, p, \gamma') \quad \text{for all } \gamma \geq \gamma'.$$

Our main interest is on the numerical approximation for the Cauchy problem:

$$-\mathcal{L}^X v - F\left(\cdot, v, Dv, D^2 v\right) = 0, \text{ on } [0, T) \times \mathbb{R}^d, \tag{12.1}$$

$$v(T, \cdot) = g, \text{ on } \in \mathbb{R}^d. \tag{12.2}$$

N. Touzi, *Optimal Stochastic Control, Stochastic Target Problems, and Backward SDE*, Fields Institute Monographs 29, DOI 10.1007/978-1-4614-4286-8_12,

12.1 Discretization

Let W be an $\mathbb{R}^d$-valued Brownian motion on a filtered probability space $(\Omega, \mathcal{F}, \mathbb{F}, \mathbb{P})$.

For a positive integer n, let $h := T/n$, $t_i = ih$, $i = 0, \ldots, n$, and consider the one step ahead Euler discretization

$$\hat{X}_h^{t,x} := x + \mu(t,x)h + \sigma(t,x)(W_{t+h} - W_t), \tag{12.3}$$

of the diffusion X corresponding to the linear operator $\mathcal{L}^X$. Our analysis does not require any existence and uniqueness result for the underlying diffusion X. However, the subsequent formal discussion assumes it in order to provide a natural justification of our numerical scheme.

Assuming that the PDE (12.1) has a classical solution, it follows from Itô's formula that

$$\mathbb{E}_{t_i,x}\left[v\left(t_{i+1}, X_{t_{i+1}}\right)\right] = v(t_i, x) + \mathbb{E}_{t_i,x}\left[\int_{t_i}^{t_{i+1}} \mathcal{L}^X v(t, X_t)\mathrm{d}t\right]$$

where we ignored the difficulties related to the local martingale part, and $\mathbb{E}_{t_i,x} := \mathbb{E}[\cdot|X_{t_i} = x]$ denotes the expectation operator conditional on $\{X_{t_i} = x\}$. Since v solves the PDE (12.1), this provides

$$v(t_i, x) = \mathbb{E}_{t_i,x}\left[v\left(t_{i+1}, X_{t_{i+1}}\right)\right] + \mathbb{E}_{t_i,x}\left[\int_{t_i}^{t_{i+1}} F(\cdot, v, Dv, D^2v)(t, X_t)\mathrm{d}t\right].$$

By approximating the Riemann integral, and replacing the process X by its Euler discretization, this suggests the following approximation:

$$v^h(T,.) := g \text{ and } v^h(t_i, x) := \mathbf{R}_{t_i}[v^h(t_{i+1},.)](x), \tag{12.4}$$

where we denoted for a function $\psi : \mathbb{R}^d \longrightarrow \mathbb{R}$ with exponential growth:

$$\mathbf{R}_t[\psi](x) := \mathbb{E}\left[\psi(\hat{X}_h^{t,x})\right] + hF\left(t, \cdot, \mathcal{D}_h\psi\right)(x), \tag{12.5}$$

with $\mathcal{D}_h\psi := \left(\mathcal{D}_h^0\psi, \mathcal{D}_h^1\psi, \mathcal{D}_h^2\psi\right)^{\mathrm{T}}$, and

$$\mathcal{D}_h^k\psi(x) := \mathbb{E}\left[D^k\psi(\hat{X}_h^{t,x})\right] \quad \text{for } k = 0, 1, 2,$$

and D^k is the kth order partial differential operator with respect to the space variable x. The differentiations in the above scheme are to be understood in the sense of distributions. This algorithm is well defined whenever g has exponential growth and F is a Lipschitz map. To see this, observe that any function with exponential growth has weak gradient and Hessian and the exponential growth is inherited at each time step from the Lipschitz property of F.

At this stage, the above backward algorithm presents the serious drawback of involving the gradient $Dv^h(t_{i+1},.)$ and the Hessian $D^2v^h(t_{i+1},.)$ in order to compute $v^h(t_i,.)$. The following result avoids this difficulty by an easy integration-by-parts argument.

Lemma 12.1. *Let $f:\mathbb{R}^d\to\mathbb{R}$ be a function with exponential growth. Then*

$$\mathbb{E}\Big[D^i f(\hat{X}_h^{t_i,x})\Big] = \mathbb{E}\Big[f(\hat{X}_h^{t_i,x})H_i^h(t_i,x)\Big] \quad \textit{for } i=1,2,$$

where

$$H_1^h = \frac{1}{h}\sigma^{-1}W_h \text{ and } H_2^h = \frac{1}{h^2}\sigma^{-1}\big(W_hW_h^{\mathrm{T}} - h\mathbf{I}_d\big)\sigma^{-1}. \tag{12.6}$$

Proof. We only provide the argument in the one-dimensional case; the extension to any dimension d is immediate. Let G be a one-dimensional Gaussian random variable with men m and variance v. Then, for any function f with exponential growth, it follows from an integration by parts that

$$\begin{aligned}\mathbb{E}[f'(G)] &= \int f'(s)\mathrm{e}^{-\frac{1}{2}\frac{(s-m)^2}{v}}\frac{\mathrm{d}s}{\sqrt{2\pi v}}\\ &= \int f(s)\frac{s-m}{v}\mathrm{e}^{-\frac{1}{2}\frac{(s-m)^2}{v}}\frac{\mathrm{d}s}{\sqrt{2\pi v}} = \mathbb{E}\left[f(G)\frac{G-m}{v}\right],\end{aligned}$$

where the remaining term in the integration-by-parts formula vanishes by the exponential growth of f. This implies the required result for $i=1$.

To obtain the result for $i=2$, we continue by integrating by parts once more:

$$\begin{aligned}\mathbb{E}[f''(G)] &= \mathbb{E}\left[f'(G)\frac{G-m}{v}\right]\\ &= \int f'(s)\frac{s-m}{v}\mathrm{e}^{-\frac{1}{2}\frac{(s-m)^2}{v}}\frac{\mathrm{d}s}{\sqrt{2\pi v}}\\ &= \int f(s)\left(-\frac{1}{v}+\left(\frac{s-m}{v}\right)^2\right)\mathrm{e}^{-\frac{1}{2}\frac{(s-m)^2}{v}}\frac{\mathrm{d}s}{\sqrt{2\pi v}}\\ &= \mathbb{E}\left[f(G)\frac{(G-m)^2-v}{v^2}\right].\end{aligned}$$

□

In the sequel, we shall denote $H_h := (1, H_1^h, H_2^h)^{\mathrm{T}}$. In view of the last lemma, we may rewrite the discretization scheme (12.4) into

$$v^h(T,.) = g \text{ and } v^h(t_i,x) = \mathbf{R}_{t_i}\left[v^h(t_{i+1},.)\right](x), \tag{12.7}$$

where

$$\mathbf{R}_{t_i}[\psi](x) = \mathbb{E}\left[\psi(\hat{X}_h^{t,x})\right] + hF\left(t, \cdot, \mathcal{D}_h\psi\right)(x)$$

and

$$\mathcal{D}_h^k\psi(x) := \mathbb{E}\left[\psi(\hat{X}_h^{t,x})H_h^k(t,x)\right] \quad \text{for } k = 0, 1, 2. \tag{12.8}$$

Observe that the choice of the drift and the diffusion coefficients μ and σ in the nonlinear PDE (12.1) is arbitrary. So far, it has been only used in order to define the underlying diffusion X. Our convergence result will however place some restrictions on the choice of the diffusion coefficient, see Remark 12.6.

Once the linear operator $\mathcal{L}^X$ is chosen in the nonlinear PDE, the above algorithm handles the remaining nonlinearity by the classical finite differences approximation. This connection with finite differences is motivated by the following formal interpretation of Lemma 12.1, where, for ease of presentation, we set $d = 1$, $\mu \equiv 0$, and $\sigma(x) \equiv 1$:

- Consider the binomial random walk approximation of the Brownian motion $\hat{W}_{t_k} := \sum_{j=1}^k w_j$, $t_k := kh$, $k \geq 1$, where $\{w_j, j \geq 1\}$ are independent random variables distributed as $\frac{1}{2}\left(\delta_{\{\sqrt{h}\}} + \delta_{\{-\sqrt{h}\}}\right)$. Then, this induces the following approximation:

$$\mathcal{D}_h^1\psi(x) := \mathbb{E}\left[\psi(X_h^{t,x})H_1^h\right] \approx \frac{\psi(x+\sqrt{h}) - \psi(x-\sqrt{h})}{2\sqrt{h}},$$

  which is the centered finite differences approximation of the gradient.
- Similarly, consider the trinomial random walk approximation $\hat{W}_{t_k} := \sum_{j=1}^k w_j$, $t_k := kh$, $k \geq 1$, where $\{w_j, j \geq 1\}$ are independent random variables distributed as $\frac{1}{6}\left(\delta_{\{\sqrt{3h}\}} + 4\delta_{\{0\}} + \delta_{\{-\sqrt{3h}\}}\right)$, so that $\mathbb{E}[w_j^n] = \mathbb{E}[W_h^n]$ for all integers $n \leq 4$. Then, this induces the following approximation:

$$\mathcal{D}_h^2\psi(x) := \mathbb{E}\left[\psi(X_h^{t,x})H_2^h\right] \approx \frac{\psi(x+\sqrt{3h}) - 2\psi(x) + \psi(x-\sqrt{3h})}{3h},$$

  which is the centered finite differences approximation of the Hessian.

In view of the above interpretation, the numerical scheme (12.7) can be viewed as a mixed Monte Carlo–Finite Differences algorithm. The Monte Carlo component of the scheme consists in the choice of an underlying diffusion process X. The finite differences component of the scheme consists in approximating the remaining nonlinearity by means of the integration-by-parts formula of Lemma 12.1.

12.2 Convergence of the Discrete-Time Approximation

The main convergence result of this section requires the following assumptions:

Assumption 12.2. *The PDE* (12.1) *has comparison for bounded functions, i.e., for any bounded upper-semicontinuous viscosity subsolution u and any bounded lower-semicontinuous viscosity supersolution v on $[0,T)\times\mathbb{R}^d$, satisfying*

$$u(T,\cdot)\le v(T,\cdot),$$

we have $u\le v$.

For our next assumption, we denote by F_r, F_p, and F_γ the partial gradients of F with respect to r, p, and γ, respectively. We also denote by F_γ^- the pseudo-inverse of the nonnegative symmetric matrix F_γ. We recall that any Lipschitz function is differentiable a.e.

Assumption 12.3. (i) *The nonlinearity F is Lipschitz continuous with respect to (x,r,p,γ) uniformly in t, and $|F(\cdot,\cdot,0,0,0)|_\infty<\infty$.*
(ii) *F is elliptic and dominated by the diffusion of the linear operator $\mathcal{L}^X$, i.e.*

$$F_\gamma\le a \quad on \quad \mathbb{R}^d\times\mathbb{R}\times\mathbb{R}^d\times\mathcal{S}_d. \tag{12.9}$$

(iii) $F_p\in\mathrm{Image}(F_\gamma)$ *and* $\left|F_p^{\mathrm{T}}F_\gamma^-F_p\right|_\infty<+\infty$.

Before commenting this assumption, we state our main convergence result.

Theorem 12.4. *Let Assumptions 12.2 and 12.3 hold true, and assume that μ,σ are Lipschitz-continuous and σ is invertible. Then for every bounded Lipschitz function g, there exists a bounded function v so that*

$$v^h\longrightarrow v \quad \textit{locally uniformly}.$$

In addition, v is the unique bounded viscosity solution of problem (12.1) *and* (12.2).

The proof of this result is reported in the Sect. 12.4. We conclude by some remarks.

Remark 12.5. Assumption 12.3(iii) is equivalent to

$$|m_F^-|_\infty<\infty \text{ where } m_F:=\min_{w\in\mathbb{R}^d}\left\{F_p\cdot w+w^{\mathrm{T}}F_\gamma w\right\}. \tag{12.10}$$

To see this observe first that F_γ is a symmetric matrix, as a consequence of the ellipticity of F. Then, any $w\in\mathbb{R}^d$ has an orthogonal decomposition $w=w_1+w_2\in\mathrm{Ker}(F_\gamma)\oplus\mathrm{Image}(F_\gamma)$, and by the nonnegativity of F_γ,

$$\begin{aligned}F_p\cdot w+w^{\mathrm{T}}F_\gamma w &= F_p\cdot w_1+F_p\cdot w_2+w_2^{\mathrm{T}}F_\gamma w_2\\ &= -\frac{1}{4}F_p^{\mathrm{T}}F_\gamma^-F_p+F_p\cdot w_1+\left|\frac{1}{2}(F_\gamma^-)^{1/2}\cdot F_p-F_\gamma^{1/2}w_2\right|^2.\end{aligned}$$

Remark 12.6. Assumption 12.3(ii) places some restrictions on the choice of the linear operator $\mathcal{L}^X$ in the nonlinear PDE (12.1). First, F is required to be uniformly elliptic, implying an upper bound on the choice of the diffusion matrix σ. Since $\sigma^2 \in \mathcal{S}_d^+$, this implies in particular that our main results do not apply to general degenerate nonlinear parabolic PDEs. Second, the diffusion of the linear operator σ is required to dominate the nonlinearity F which places implicitly a lower bound on the choice of the diffusion σ.

Example 12.7. Let us consider the nonlinear PDE in the one-dimensional case $-\frac{\partial v}{\partial t} - \frac{1}{2}\left(a^2 v_{xx}^+ - b^2 v_{xx}^-\right)$ where $0 < b < a$ are given constants. Then if we restrict the choice of the diffusion to be constant, it follows from Assumption 12.3 that $\frac{1}{3}a^2 \le \sigma^2 \le b^2$, which implies that $a^2 \le 3b^2$. If the parameters a and b do not satisfy the latter condition, then the diffusion σ has to be chosen to be state and time dependent.

Remark 12.8. Under the boundedness condition on the coefficients μ and σ, the restriction to a bounded terminal data g in Theorem 12.4 can be relaxed by an immediate change of variable. Let g be a function with α−exponential growth for some $\alpha > 0$. Fix some $M > 0$, and let ρ be an arbitrary smooth positive function with

$$\rho(x) = \mathrm{e}^{\alpha|x|} \quad \text{for } |x| \ge M,$$

so that both $\rho(x)^{-1}\nabla\rho(x)$ and $\rho(x)^{-1}\nabla^2\rho(x)$ are bounded. Let

$$u(t,x) := \rho(x)^{-1}v(t,x) \quad \text{for } (t,x) \in [0,T] \times \mathbb{R}^d.$$

Then, the nonlinear PDE problem (12.1) and (12.2) satisfied by v converts into the following nonlinear PDE for u:

$$\begin{aligned} -\mathcal{L}^X u - \tilde{F}\left(\cdot, u, Du, D^2u\right) &= 0 \ \text{ on } [0,T) \times \mathbb{R}^d \\ v(T,\cdot) = \tilde{g} &:= \rho^{-1}g \ \text{ on } \mathbb{R}^d, \end{aligned} \tag{12.11}$$

where

$$\begin{aligned} \tilde{F}(t,x,r,p,\gamma) := {} & r\mu(x)\cdot\rho^{-1}\nabla\rho + \frac{1}{2}\mathrm{Tr}\left[a(x)\left(r\rho^{-1}\nabla^2\rho + 2p\rho^{-1}\nabla\rho^{\mathrm{T}}\right)\right] \\ & + \rho^{-1}F\left(t,x,r\rho, r\nabla\rho + p\rho, r\nabla^2\rho + 2p\nabla\rho^{\mathrm{T}} + \rho\gamma\right). \end{aligned}$$

Recall that the coefficients μ and σ are assumed to be bounded. Then, it is easy to see that $\tilde{F}$ satisfies the same conditions as F. Since $\tilde{g}$ is bounded, the convergence Theorem 12.4 applies to the nonlinear PDE (12.11).

12.3 Consistency, Monotonicity and Stability

The proof of Theorem 12.4 is based on the monotone schemes method of Barles and Souganidis [4] which exploits the stability properties of viscosity solutions. The monotone schemes method requires three conditions: consistency, monotonicity, and stability that we now state in the context of backward scheme (12.7).

To emphasize on the dependence on the small parameter h in this section, we will use the notation

$$\mathbf{T}_h[\varphi](t,x) := \mathbf{R}_t[\varphi(t+h,.)](x) \quad \text{for all } \varphi : \mathbb{R}_+ \times \mathbb{R}^d \longrightarrow \mathbb{R}.$$

Lemma 12.9 (Consistency). *Let φ be a smooth function with bounded derivatives. Then for all $(t,x) \in [0,T] \times \mathbb{R}^d$*

$$\lim_{\substack{(t',x') \to (t,x) \\ (h,c) \to (0,0) \\ t'+h \le T}} \frac{\big([c+\varphi] - \mathbf{T}_h[c+\varphi]\big)(t',x')}{h} = -\left(\mathcal{L}^X \varphi + F(\cdot,\varphi,D\varphi,D^2\varphi)\right)(t,x).$$

The proof is a straightforward application of Itô's formula, and is omitted.

Lemma 12.10 (Monotonicity). *Let $\varphi, \psi : [0,T] \times \mathbb{R}^d \longrightarrow \mathbb{R}$ be two Lipschitz functions with $\varphi \le \psi$. Then*

$$\mathbf{T}_h[\varphi](t,x) \le \mathbf{T}_h[\psi](t,x) + Ch\, \mathbb{E}\left[(\psi - \varphi)(t+h, \hat{X}_h^{t,x})\right] \quad \textit{for some } C > 0$$

depending only on the constant m_F in (12.10).

Proof. By Lemma 12.1, the operator $\mathbf{T}_h$ can be written as

$$\mathbf{T}_h[\psi](t,x) = \mathbb{E}\left[\psi(\hat{X}_h^{t,x})\right] + hF\left(t,x,\mathbb{E}\left[\psi(\hat{X}_h^{t,x})H_h(t,x)\right]\right).$$

Let $f := \psi - \varphi \ge 0$ where φ and ψ are as in the statement of the lemma. Let F_τ denote the partial gradient with respect to $\tau = (r,p,\gamma)$. By the mean value theorem

$$\begin{aligned}\mathbf{T}_h[\psi](t,x) - \mathbf{T}_h[\varphi](t,x) &= \mathbb{E}\left[f(\hat{X}_h^{t,x})\right] + hF_\tau(\theta)\cdot\mathcal{D}_h f(\hat{X}_h^{t,x}) \\ &= \mathbb{E}\left[f(\hat{X}_h^{t,x})\left(1 + hF_\tau(\theta)\cdot H_h(t,x)\right)\right],\end{aligned}$$

for some $\theta = (t,x,\bar{r},\bar{p},\bar{\gamma})$. By the definition of $H_h(t,x)$

$$\begin{aligned}&\mathbf{T}_h[\psi] - \mathbf{T}_h[\varphi] \\ &\quad = \mathbb{E}\left[f(\hat{X}_h^{t,x})\left(1 + hF_r + F_p.\sigma^{-1}W_h + h^{-1}F_\gamma\cdot\sigma^{-1}(W_hW_h^{\mathrm{T}} - hI)\sigma^{-1}\right)\right],\end{aligned}$$

where the dependence on θ and x has been omitted for notational simplicity. Since $F_\gamma \le a$ by Assumption 12.3, we have $1 - a^{-1} \cdot F_\gamma \ge 0$ and therefore

$$\mathbf{T}_h[\psi] - \mathbf{T}_h[\varphi] \ge \mathbb{E}\Big[f(\hat{X}_h^{t,x})\left(hF_r + F_p.\sigma^{-1}W_h + h^{-1}F_\gamma \cdot \sigma^{-1}W_hW_h^{\mathrm{T}}\sigma^{-1}\right)\Big]$$
$$= \mathbb{E}\left[f(\hat{X}_h^{t,x})\left(hF_r + hF_p.\sigma^{-1}\frac{W_h}{h} + hF_\gamma \cdot \sigma^{-1}\frac{W_hW_h^{\mathrm{T}}}{h^2}\sigma^{-1}\right)\right].$$

Recall the function m_F defined in (12.10). Under Assumption 12.3, it follows from Remark 12.5 that $K := |m_F^-|_\infty < \infty$. Then

$$F_p.\sigma^{-1}\frac{W_h}{h} + hF_\gamma \cdot \sigma^{-1}\frac{W_hW_h^{\mathrm{T}}}{h^2}\sigma^{-1} \ge -K,$$

and therefore:

$$\mathbf{T}_h[\psi] - \mathbf{T}_h[\varphi] \ge \mathbb{E}\Big[f(\hat{X}_h^{t,x})\,(hF_r - hK)\Big] \ \ge\ -C'h\mathbb{E}\Big[f(\hat{X}_h^{t,x})\Big]$$

for some constant $C > 0$, where the last inequality follows from (12.10). □

Lemma 12.11 (Stability). *Let $\varphi, \psi : [0, T] \times \mathbb{R}^d \longrightarrow \mathbb{R}$ be two $\mathbb{L}^\infty$–bounded functions. Then there exists a constant $C > 0$ such that*

$$\big|\mathbf{T}_h[\varphi] - \mathbf{T}_h[\psi]\big|_\infty \le |\varphi - \psi|_\infty(1 + Ch).$$

In particular, if g is $\mathbb{L}^\infty$–bounded, then the family $(v^h)_h$ defined in (12.7) is $\mathbb{L}^\infty$–bounded, uniformly in h.

Proof. Let $f := \varphi - \psi$. Arguing as in the previous proof, we see that

$$\mathbf{T}_h[\varphi] - \mathbf{T}_h[\psi] = \mathbb{E}\left[f(\hat{X}_h)\left(1 - a^{-1} \cdot F_\gamma + h|A_h|^2 + hF_r - \frac{h}{4}F_p^{\mathrm{T}}F_\gamma^- F_p\right)\right].$$

where

$$A_h = \frac{1}{2}(F_\gamma^-)^{1/2}F_p - F_\gamma^{1/2}\sigma^{-1}\frac{W_h}{h}.$$

Since $1 - \mathrm{Tr}[a^{-1}F_\gamma] \ge 0$, $|F_r|_\infty < \infty$, and $|F_p^{\mathrm{T}}F_\gamma^- F_p|_\infty < \infty$ by Assumption 12.3, it follows that

$$|\mathbf{T}_h[\varphi] - \mathbf{T}_h[\psi]|_\infty \le |f|_\infty\left(1 - a^{-1} \cdot F_\gamma + h\mathbb{E}[|A_h|^2] + Ch\right).$$

But, $\mathbb{E}[|A_h|^2] = \frac{h}{4}F_p^{\mathrm{T}}F_\gamma^- F_p + a^{-1} \cdot F_\gamma$. Therefore, using again Assumption 12.3, we see that

$$|\mathbf{T}_h[\varphi] - \mathbf{T}_h[\psi]|_\infty \le |f|_\infty \left(1 + \frac{h}{4} F_p^{\mathrm{T}} F_\gamma^{-} F_p + Ch\right) \le |f|_\infty (1 + \bar{C}h).$$

To prove that the family $(v^h)_h$ is bounded, we proceed by backward induction. By the assumption of the lemma $v^h(T,.) = g$ is $\mathbb{L}^\infty$–bounded. We next fix some $i < n$ and we assume that $|v^h(t_j,.)|_\infty \le C_j$ for every $i + 1 \le j \le n - 1$. Proceeding as in the proof of Lemma 12.10 with $\varphi \equiv v^h(t_{i+1},.)$ and $\psi \equiv 0$, we see that

$$\left|v^h(t_i,.)\right|_\infty \le h\,|F(t,x,0,0,0)| + C_{i+1}(1 + Ch).$$

Since $F(t,x,0,0,0)$ is bounded by Assumption 12.3, it follows from the discrete Gronwall inequality that $|v^h(t_i,.)|_\infty \le C\mathrm{e}^{CT}$ for some constant C independent of h. □

12.4 The Barles–Souganidis Monotone Scheme

This section is dedicated to the proof of Theorem 12.4. We emphasize on the fact that the subsequent argument applies to any numerical scheme which satisfies the consistency, monotonicity, and stability properties. In the present situation, we also need to prove a technical result concerning the limiting behavior of the boundary condition at T. This will be needed in order to use the comparison result which is assumed to hold for the equation. The statement and its proof are collected in Lemma 12.12.

Proof of Theorem 12.4. 1. By the stability property of Lemma 12.11, it follows that the relaxed semicontinuous envelopes

$$\underline{v}(t,x) := \liminf_{(h,t',x')\to(0,t,x)} v^h(t',x') \text{ and } \overline{v}(t,x) := \limsup_{(h,t',x')\to(0,t,x)} v^h(t',x')$$

are bounded. We shall prove in Step 2 that $\underline{v}$ and $\overline{v}$ are viscosity supersolution and subsolution, respectively. The final ingredient is reported in Lemma 12.12 which states that $\underline{v}(T,.) = \overline{v}(T,.)$. Then, the proof is completed by appealing to the comparison property of Assumption 12.2.

2. We only prove that $\underline{v}$ is a viscosity supersolution of (12.1). The proof of the viscosity subsolution property of $\overline{v}$ follows exactly the same line of argument. Let $(t_0,x_0) \in [0,T) \times \mathbb{R}^d$ and $\varphi \in C^2\left([0,T] \times \mathbb{R}^d\right)$ be such that

$$0 = (\underline{v} - \varphi)(t_0,x_0) = (\text{strict}) \min_{[0,T]\times\mathbb{R}^d} (\underline{v} - \varphi). \tag{12.12}$$

Since v^h is uniformly bounded in h, we may assume without loss of generality that φ is bounded. Let $(h_n, t_n, x_n)_n$ be a sequence such that

$$h_n \to 0,\ (t_n, x_n) \to (t_0, x_0),\ \text{and } v^{h_n}(t_n, x_n) \longrightarrow \underline{v}(t_0, x_0). \tag{12.13}$$

For a positive scalar r with $2r < T - t_0$, we denote by $\overline{B}_r(t_n, x_n)$ the ball of radius r centered at (t_n, x_n), and we introduce

$$\delta_n := (v_*^{h_n} - \varphi)(\hat{t}_n, \hat{x}_n) = \min_{\overline{B}_r(t_n, x_n)} (v_*^{h_n} - \varphi), \tag{12.14}$$

where $v_*^{h_n}$ is the lower-semicontinuous envelope of v^{h_n}. We claim that

$$\delta_n \longrightarrow 0 \text{ and } (\hat{t}_n, \hat{x}_n) \longrightarrow (t_0, x_0). \tag{12.15}$$

This claim is proved in Step 3 . By the definition of $v_*^{h_n}$, we may find a sequence $(\hat{t}'_n, \hat{x}'_n)_{n\geq 1}$ converging to (t_0, x_0), such that

$$|v^{h_n}(\hat{t}'_n, \hat{x}'_n) - v_*^{h_n}(\hat{t}_n, \hat{x}_n)| \leq h_n^2 \text{ and } |\varphi(\hat{t}'_n, \hat{x}'_n) - \varphi(\hat{t}_n, \hat{x}_n)| \leq h_n^2. \tag{12.16}$$

By (12.14), (12.16), and the definition of the functions v^h in (12.7), we have

$$\begin{aligned}
2h_n^2 + \delta_n + \varphi(\hat{t}'_n, \hat{x}'_n) &\geq h_n^2 + \delta_n + \varphi(\hat{t}_n, \hat{x}_n)\\
&= h_n^2 + v_*^{h_n}(\hat{t}_n, \hat{x}_n)\\
&\geq v^{h_n}(\hat{t}'_n, \hat{x}'_n)\\
&= \mathbf{T}_{h_n}[v^{h_n}](\hat{t}'_n, \hat{x}'_n)\\
&\geq \mathbf{T}_{h_n}[\varphi^{h_n} + \delta_n](\hat{t}'_n, \hat{x}'_n)\\
&\quad + C h_n \mathbb{E}\left[(v^{h_n} - \varphi - \delta_n)\left(\hat{X}_{h_n}^{\hat{t}'_n, \hat{x}'_n}\right)\right],
\end{aligned}$$

where the last inequality follows from (12.14) and the monotonicity property of Lemma 12.10. Dividing by h_n, the extremes of this inequality provide

$$\frac{\delta_n + \varphi(\hat{t}'_n, \hat{x}'_n) - \mathbf{T}_{h_n}[\varphi^{h_n} + \delta_n](\hat{t}'_n, \hat{x}'_n)}{h_n} \geq C\mathbb{E}\left[(u^{h_n} - \varphi - \delta_n)\left(\hat{X}_{h_n}^{\hat{t}'_n, \hat{x}'_n}\right)\right].$$

We now send n to infinity. The right hand side converges to zero by (12.13), (12.15), and the dominated convergence theorem. For the left hand-side term, we use the consistency result of Lemma 12.9. This leads to

$$\big(-\mathcal{L}^X\varphi - F(., \varphi, D\varphi, D^2\varphi)\big)(t_0, x_0) \geq 0,$$

as required.

3. We now prove Claim (12.15). Since $(\hat{t}_n, \hat{x}_n)_n$ is a bounded sequence, we may extract a subsequence, still named $(\hat{t}_n, \hat{x}_n)_n$, converging to some $(\hat{t}, \hat{x})$. Then:

$$
\begin{aligned}
0 &= (\underline{v} - \varphi)(t_0, x_0) \\
&= \lim_{n\to\infty} (v^{h_n} - \varphi)(t_n, x_n) \\
&\geq \limsup_{n\to\infty} (v_*^{h_n} - \varphi)(t_n, x_n) \\
&\geq \limsup_{n\to\infty} (v_*^{h_n} - \varphi)(\hat{t}_n, \hat{x}_n) \\
&\geq \liminf_{n\to\infty} (v_*^{h_n} - \varphi)(\hat{t}_n, \hat{x}_n) \\
&\geq (\underline{v} - \varphi)(\hat{t}, \hat{x}).
\end{aligned}
$$

Since (t_0, x_0) is a strict minimizer of the difference $(\underline{v} - \varphi)$, this implies (12.15). □

The following result is needed in order to use the comparison property of Assumption 12.2. We shall not report its long technical proof, see [19].

Lemma 12.12. *The function v^h is Lipschitiz in x, $1/2$–Hölder continuous in t, uniformly in h, and for all $x \in \mathbb{R}^d$, we have*

$$|v^h(t,x) - g(x)| \leq C(T-t)^{\frac{1}{2}}.$$

Chapter 13
Introduction to Finite Differences Methods

by Agnès Tourin*

In this lecture, I discuss the practical aspects of designing Finite Difference methods for Hamilton–Jacobi–Bellman equations of parabolic type arising in quantitative finance. The approach is based on the very powerful and simple framework developed by Barles– Souganidis [4], see the review of the previous chapter. The key property here is the monotonicity which guarantees that the scheme satisfies the same ellipticity condition as the HJB operator. I will provide a number of examples of monotone schemes in these notes. In practice, pure finite difference schemes are only useful in 1, 2, or at most 3 spatial dimensions. One of their merits is to be quite simple and easy to implement. Also, as shown in the previous chapter, they can also be combined with Monte Carlo methods to solve nonlinear parabolic PDEs.

Such approximations are now fairly standard, and you will find many interesting examples available in the literature. For instance, I suggest the articles on the subject by Forsyth (see [33, 41, 43]). There is also a classical book written by Kushner and Paul Dupuis [28] on numerical methods for stochastic control problems. Finally, for a basic introduction to finite difference methods for linear parabolic PDEs, I recommend the book by Thomas [40].

13.1 Overview of the Barles–Souganidis Framework

Consider the parabolic PDE

$$u_t - F(t, x, u, Du, D^2u) = 0 \text{ in } (0, T] \times \mathbb{R}^N, \tag{13.1}$$

$$u(0, x) = u_0(x) \text{ in } \mathbb{R}^N, \tag{13.2}$$

*Fields Institute Research Immersion Fellow, 2010

N. Touzi, *Optimal Stochastic Control, Stochastic Target Problems, and Backward SDE*, Fields Institute Monographs 29, DOI 10.1007/978-1-4614-4286-8_13,

where F is elliptic

$$F(t,x,u,p,A) \geq F(t,x,u,p,B), \quad \text{if } A \geq B.$$

For the sake of simplicity, we assume that u_0 is bounded in $\mathbb{R}^N$.

The main application we have in mind is, for instance, to an operator F coming from a stochastic control problem:

$$F(t,x,r,p,X) = \sup_{\alpha \in \mathcal{A}} \{-\mathrm{Tr}[a^\alpha(t,x)X] - b^\alpha(t,x)p - c^\alpha(t,x)r - f^\alpha(t,x)\},$$

where $a^\alpha = \frac{1}{2}\sigma^\alpha \sigma^{\alpha T}$.

Typically, the set of control $\mathcal{A}$ is compact or finite, all the coefficients in the equations are bounded and Lipschitz continuous in x, Hölder with coefficient $\frac{1}{2}$ in t and all the bounds are independent of α. Then the unique viscosity solution u of (13.1) is a bounded and Lipschitz continuous function and is the solution of the underlying stochastic control problem. The ideas, concepts, and techniques actually apply to a broader range of optimal control problems. In particular, you can adapt the techniques to handle different situations, even possibly treat some delicate singular control problems.

In the previous chapter, our convergence result required the technical, Lemma 12.12 in order for the comparison result to apply. However, an easier statement of the Barles–Souganidis method can be obtained at the price of assuming a stronger comparison result in the following sense.

Definition 13.1. We say that the problem (13.1) and (13.2) satisfies the strong comparison principle for bounded solution if for all bounded functions $u \in$ USC and $v \in$ LSC such that

- u (resp. v) is a viscosity subsolution (resp. supersolution) of (13.1) on $(0,T] \times \mathbb{R}^N$.
- The boundary condition holds in the viscosity sense

$$\max\{u_t - F(.,u,Du,D^2u), u - u_0\} \geq 0 \text{ on } \{0\} \times \mathbb{R}^N$$
$$\min\{u_t - F(.,u,Du,D^2u), u - u_0\} \leq 0 \text{ on } \{0\} \times \mathbb{R}^N;$$

we have $u \leq v$ on $[0,T] \times \mathbb{R}^N$.

Under the strong comparison principle, any monotonic stable and consistent scheme achieves convergence, and there is no need to analyze the behavior of the scheme near the boundary.

The aim is to build an approximation scheme which preserves the ellipticity. This discrete ellipticity property is called *monotonicity*. The monotonicity, together with the consistency of the scheme and some regularity, ensures its convergence to the unique viscosity solution of the PDE (13.1), (13.2). It is worth insisting on the fact that if the scheme is not monotone, it may fail to converge to the correct solution (see [33] for an example)! We present the theory rather informally, and we refer to the original articles for more details. The general concepts and machinery apply to a wide range of equations, but the reader needs to be aware that each PDE has its own peculiarities and that, in practice, the techniques must be tailored to each particular application.

A numerical scheme is an equation of the following form:

$$S(h,t,x,u_h(t,x),[u_h]_{t,x}) = 0 \quad \text{for } (t,x) \text{ in } \mathcal{G}_h \backslash \{t=0\} \tag{13.3}$$

$$u_h(0,x) = u_0(x) \text{ in } \mathcal{G}_h \cap \{t=0\} \tag{13.4}$$

where $h = (\Delta t, \Delta x)$, $\mathcal{G}_h = \Delta t\{0,1,\ldots,n_T\} \times \Delta x Z^N$ is the grid, u_h stands for the approximation of u on the grid, $u_h(t,x)$ is the approximation u_h at the point (t,x), and $[u_h]_{t,x}$ represents the value of u_h at other points than (t,x). Note that u_h can be both interpreted as a function defined at the grid points only or on the whole space. Indeed if one knows the value of u_h on the mesh, a continuous version of u_h can be constructed by linear interpolation.

The first and crucial condition in the Barles–Souganidis framework is:

Monotonicity $S(h,t,x,r,u) \geq S(h,t,x,r,v)$ whenever $u \leq v$.

The monotonicity assumption can be weakened. This was indeed the case in the previous chapter. We only need it to hold approximately, with a margin of error that vanishes to 0 as h goes to 0.

Consistency *For every smooth function $\phi(t,x)$,*

$$\lim_{h\to 0,(n\Delta t, i\Delta x)\to(t,x),c\to 0} S(h, n\Delta t, i\Delta x, \Phi(t,x)+c, [\Phi(t,x)+c]_{t,x})$$
$$= \Phi_t + F(t,x,\Phi(t,x), D\Phi, D^2\Phi).$$

The final condition is:

Stability *For every $h > 0$, the scheme has a solution u_h which is uniformly bounded independently of h.*

Theorem 13.2. *Assume that the problem* (13.1) *and* (13.2) *satisfies the strong comparison principle for bounded functions. Assume further that the scheme (13.3), (13.4) satisfies the consistency, monotonicity, and stability properties. Then, its solution u_h converges locally uniformly to the unique viscosity solution of (13.1), (13.2).*

13.2 First Examples

13.2.1 The Heat Equation: The Classic Explicit and Implicit Schemes

First, let me recall the classic explicit and implicit schemes for the heat equation:

$$u_t - u_{xx} = 0 \text{in} (0,T] \times \mathbb{R}. \tag{13.5}$$

We claim that for any choice of the time step, the implicit scheme is *monotone*. In order to verify that claim, let us rewrite the implicit scheme using the notation S:

$$S(\Delta t, \Delta x, (n+1)\Delta T, i\Delta x, u_i^{n+1}, [u_{i-1}^{n+1}, u_i^n, u_{i+1}^{n+1}])$$
$$= \frac{u_i^{n+1} - u_i^n}{\Delta t} - \frac{u_{i+1}^{n+1} + u_{i-1}^{n+1} - 2u_i^{n+1}}{\Delta X^2}.$$

Since S is decreasing in u_i^n, u_{i+1}^{n+1}, and u_{i-1}^{n+1}, the implicit scheme is unconditionally *monotone*.

13.2.2 The Black–Scholes–Merton PDE

The price of a European call $u(t,x)$ satisfies the degenerate linear PDE

$$u_t + ru - \tfrac{1}{2}\sigma^2 x^2 u_{xx} - rxu_x = 0 \text{ in } (0,T] \times (0,+\infty)$$
$$u(0,x) = (x-K)^+.$$

The Black–Scholes–Merton PDE is linear and its elliptic operator is degenerate. The first derivative u_x can be easily approximated in a monotone way using a forward finite difference

$$-rxu_x \approx -rx_i \frac{u_{i+1}^{n+1} - u_i^{n+1}}{\Delta x}.$$

One can, for instance, implement the implicit scheme

$$S(\Delta t, \Delta x, (n+1)\Delta T, i\Delta x, u_i^{n+1}, [u_{i-1}^{n+1}, u_i^n, u_{i+1}^{n+1}])$$
$$= \frac{u_i^{n+1} - u_i^n}{\Delta t} + ru_i^{n+1} - \frac{1}{2}(i\Delta x)^2 \frac{u_{i+1}^{n+1} + u_{i-1}^{n+1} - 2u_i^{n+1}}{\Delta X^2} - ri\Delta x \frac{u_{i+1}^{n+1} - u_i^{n+1}}{\Delta x}.$$

13.3 A Nonlinear Example: The Passport Option

13.3.1 Problem Formulation

It is an interesting example of a one-dimensional nonlinear HJB equation. I present only briefly the underlying model here and refer to the article [41] for more details and references. A passport option is an option on the balance of a trading account. The holder can trade an underlying asset S over a finite time horizon $[0,T]$. At maturity, the holder keeps any net gain, while the writer bears any loss. The number of shares q of stock held is bounded by a given number C. Without any loss of generality, this number is commonly assumed to be 1 (the problem can be solved in full generality by using the appropriate scaling). The stock S follows a geometric Brownian motion with drift μ and volatility σ, r is the risk-free interest rate, γ is the dividend rate, r_t is the interest rate for the trading account, and r_c is its cost of carry rate. The option price $V(t, S, W)$, which depends on S and on the accumulated wealth W in the trading account, solves the PDE

$$-V_t + rV - (r - \gamma)SV_S - \sup_{|q|\leq 1}\left\{ -((\gamma - r + r_c)qS - r_t W)V_W + \frac{1}{2}\sigma^2 S^2 (V_{SS} + 2qV_{SW} + q^2 V_{WW})\right\} = 0$$

$$V(T, S, W) = \max(W, 0).$$

Next, one can reduce this problem to a one-dimensional equation by introducing the variable $x = W/S$ and the new function u satisfying $V(T - t, S, W) = Su(t, x)$. The PDE for u then reads

$$u_t + \gamma u - \sup_{|q|\leq 1}\{((r - \gamma - r_c)q - (r - \gamma - r_t)x)u_x + \frac{1}{2}\sigma^2 (x - q)^2 u_{xx}\}$$

$$u(0, x) = \max(x, 0).$$

Note that, in this example, the solution is no longer bounded but grows at most linearly at infinity. The Barles and Souganidis [4] framework can be slightly modified to accommodate the linear growth of the value function at infinity.

When the payoff is convex, it is easy to see that the optimal value for q is either $+1$ or -1. When the payoff is no longer convex, the supremum may be achieved inside the interval at $q^* = x - \frac{(r-\gamma-r_c)u_x}{\sigma^2 u_{xx}}$. For simplicity, we consider only the convex case.

13.3.2 Finite Difference Approximation

To simplify further, we focus on a simple case: we assume that $r - \gamma - r_t = 0$ and $r - \gamma - r_c < 0$. This equation is still fairly difficult to solve because the approximation scheme must depend on the control q.

$$u_t + \gamma u - \max\left\{(r - \gamma - r_c)u_x + \frac{1}{2}\sigma^2 (x - 1)^2 u_{xx}\, ,\; -(r - \gamma - r_c)u_x + \frac{1}{2}\sigma^2 (x + 1)^2 u_{xx}\right\}$$

$$u(0, x) = \max(x, 0)$$

One can easily construct an explicit monotone scheme by using the appropriate forward or backward finite difference for the first partial derivative. Often, this type of scheme is called "upwind" because you move along the direction prescribed by the deterministic dynamics $b(x,\alpha^*)$ corresponding to the optimal control α^* and pick the corresponding neighbor. For instance, for the passport option, the dynamics is

$$\begin{aligned}\text{For } q^* &= 1, b^{\alpha^*}(t,x) = q^*(r-\gamma-r_c) = (r-\gamma-r_c) < 0\\ \text{For } q^* &= -1, b^{\alpha^*}(t,x) = -(r-\gamma-r_c) > 0,\end{aligned}$$

and the corresponding upwind finite differences are

$$\begin{aligned}\text{For } q^* &= 1, u_x \approx D^- u_i^n\\ \text{For } q^* &= -1, u_x \approx D^+ u_i^n,\end{aligned}$$

where we used the standard notations

$$D^- u_i^n = \frac{u_i^n - u_{i-1}^n}{\Delta x},\ D^+ u_i^n = \frac{u_{i+1}^n - u_i^n}{\Delta x}.$$

Then the scheme reads

$$\frac{u_i^{n+1} - u_i^n}{\Delta t} + \gamma u_i^n - \max\left\{(r-\gamma-r_c)\frac{u_i^n - u_{i-1}^n}{\Delta x} + \frac{1}{2}\sigma^2(x_i-1)^2\frac{u_{i+1}^n + u_{i-1}^n - 2u_i^n}{\Delta x^2},\right.$$
$$\left.-(r-\gamma-r_c)\frac{u_{i+1}^n - u_i^n}{\Delta x} + \frac{1}{2}\sigma^2(x_i+1)^2\frac{u_{i+1}^n + u_{i-1}^n - 2u_i^n}{\Delta x^2}\right\} = 0.$$

This scheme clearly satisfies the monotonicity assumption under the CFL condition

$$\Delta t \le \frac{1}{\gamma + \frac{|r-\gamma-r_c|}{\Delta x} + \frac{\sigma^2 \max\{\max_i\{i\Delta x - 1\}^2, \max_i\{i\Delta x+1\}^2\}}{\Delta x^2}}.$$

Approximating the first spatial derivative by the classic centered finite difference, i.e., $u_x \approx \frac{u_{i+1}^n - u_{i-1}^n}{2\Delta x}$, would not yield a monotone scheme here.

Note that this condition is very restrictive. First of all, as expected, Δt has to be of order Δx^2. Furthermore, Δt also depends on the size of the grid through the terms $(i\Delta x - 1)^2, (i\Delta x - 1)^2$ and even approaches 0 as the size of the domain goes to infinity. In this situation, we renounce using the above explicit scheme and replace it by the fully implicit upwind scheme which is unconditionally monotone.

$$\frac{u_i^{n+1}-u_i^n}{\Delta t}+\gamma u_i^{n+1}-\max\left\{(r-\gamma-r_c)\frac{u_i^{n+1}-u_{i-1}^{n+1}}{\Delta x}\right.$$

$$+\frac{1}{2}\sigma^2(x_i-1)^2\frac{u_{i+1}^{n+1}+u_{i-1}^{n+1}-2u_i^{n+1}}{\Delta x^2},-(r-\gamma-r_c)\frac{u_{i+1}^{n+1}-u_i^{n+1}}{\Delta x}$$

$$\left.+\frac{1}{2}\sigma^2(x_i+1)^2\frac{u_{i+1}^{n+1}+u_{i-1}^{n+1}-2u_i^{n+1}}{\Delta x^2}\right\}=0.$$

Inverting the above scheme is challenging because it depends on the control. This can be done using the classic iterative Howard algorithm which we describe below in a general setting. However, it may be time-consuming to compute the solution of a nonlinear finite difference scheme, i.e., invert an implicit scheme using an iterative method.

13.3.3 Howard Algorithm

We denote by u_h^n, u_h^{n+1} the approximations at time n and $n+1$. We can rewrite the scheme that we need to invert as

$$\min_\alpha\{A_h^\alpha u_h^{n+1}-B_h^\alpha u_h^n\}=0.$$

Step 0: Start with an initial value for the control α_0. Compute the solution v_h^0 of $A_h^{\alpha_0}w-B_h^{\alpha_0}u_h^n=0$.

Step $k\to k+1$: Given v_h^k, find α_{k+1} minimizing $A_h^\alpha v_h^k-B_h^\alpha u_h^n$. Then compute the solution v_h^{k+1} of $A_h^{\alpha_{k+1}}w-B_h^{\alpha_{k+1}}u_h^n=0$.

Final step: *If* $|v_h^{k+1}-v_h^k|<\epsilon$, then set $u_h^{n+1}=v_h^{k+1}$.

13.4 The Bonnans–Zidani [7] Approximation

Sometimes, for a given problem, it is very difficult or even impossible to find a monotone scheme. Rewriting the PDE in terms of directional derivatives instead of partial derivatives can be extremely useful. For example, in two spatial dimensions, a naive discretization of the partial derivative v_{xy} may fail to be monotone. In fact, approximating second-order operators with crossed derivatives in a monotone way is not easy. You actually need to be able to interpret your second-order term as a directional derivative (of a linear combination of directional derivatives) and approximate each directional derivative by the adequate finite difference. In other words, you need to "move in the right direction" in order to preserve the elliptic structure of the operator.

Here is, for instance, a *naive approximation* of v_{xy} (assume $\Delta x = \Delta y$):

$$v_{xy} \approx \frac{v_{i+1,j+1} + v_{i-1,j-1} - v_{i+1,j-1} - v_{i-1,j+1}}{4\Delta x^2}.$$

It is consistent but clearly not monotone (the terms $v_{i-1,j+1}, v_{i+1,j-1}$ have the wrong sign).

Instead, let us follow Bonnans and Zidani [7]. Consider the second-order derivative

$$L^\alpha \Phi(t,x) = tr(a^\alpha(t,x) D^2 \Phi(t,x))$$

and assume that the coefficients a^α admit the decomposition

$$a^\alpha(t,x) = \sum_\beta \overline{a}^\alpha_\beta \beta\beta^T,$$

where the coefficients $\overline{a}^\alpha_\beta$ are positive. The operator can then be expressed in terms of the directional derivatives $D^2_\beta = tr[\beta\beta^T D^2]$

$$L^\alpha \Phi(t,x) = \sum_\beta \overline{a}^\alpha_\beta(t,x) D^2_\beta \Phi(t,x).$$

Finally, we can use the consistent and monotone approximation for each directional derivative

$$D^2_\beta v(t,x) \approx \frac{v(t,x+\beta\Delta x) + v(t,x-\beta\Delta x) - 2v(t,x)}{|\beta|^2 \Delta x^2}.$$

In practice, if the points $x + \beta\Delta x$, $x - \beta\Delta x$ are not on the grid, you need to estimate the value of v at these points by simple linear interpolation between two grid points. Of course, you have to make sure that the interpolation procedure preserves the monotonicity of the approximation.

Comments

- In all the above examples, I only consider the immediate neighbors of a given point $((n+1)\Delta t, i\Delta x)$, namely, $(n\Delta t, i\Delta x)$, $(n\Delta t, (i-1)\Delta x)$, $(n\Delta t, (i+1)\Delta x)$, $((n+1)\Delta t, (i-1)\Delta x$, and $(n+1)\Delta t, (i+1)\Delta x)$. Sometimes, it is worth considering a larger neighborhood and picking neighbors located further away from $((n+1)\Delta t, i\Delta x)$. It is particularly useful for the discretization of a transport term with a high speed, when information "travels fast."
- The theoretical accuracy of a monotone finite difference scheme is quite low. The Barles–Jakobsen theory [3] predicts a typical rate of $1/5$ ($|h|^{\frac{1}{5}}$ where $h = \sqrt{\Delta x^2 + \Delta t}$ and an optimal rate of $1/2$. Sometimes, higher rates are reported in practice (first order).

13.5 Working in a Finite Domain

When one implements a numerical scheme, one cannot work on the whole space and must instead work on a finite grid. Consequently, one has to impose some extra boundary conditions at the edges of the grid. This creates an additional source of error and even sometimes instabilities. Indeed, when the behavior at infinity is not known, imposing an overestimated boundary condition may cause the computed solution to blow up. If the behavior of the solution at infinity is known, it is then relatively easy to come up with a reasonable boundary condition. Next, one can try to prove that the extra error introduced is confined within a boundary layer or more precisely decreases exponentially as a function of the distance to the boundary (see [5] for a result in this direction). Also, one can perform experiments to ensure that these artificial boundary conditions do not affect the accuracy of the results, by increasing the size of the domain and checking that the first six significant digits of the computed solution are not affected.

13.6 Variational Inequalities and Splitting Methods

13.6.1 The American Option

This is the easiest example of variational inequalities arising in finance, and it gives the opportunity to introduce splitting methods. We look at the simplified VI: $u(t,x)$ solves

$$\max(u_t - u_{xx}, u - \psi(t,x)) = 0 \text{ in } (0,T] \times \mathbb{R} \tag{13.7}$$

$$u(0,x) = u_0(x). \tag{13.8}$$

This PDE can be approximated using the following semi-discretized scheme

Step 1. Given u^n, solve the heat equation

$$w_t - w_{xx} = 0 \text{ in } (n\Delta t, (n+1)\Delta t] \times \mathbb{R} \tag{13.9}$$

$$w(n\Delta t, x) = u^n(x), \tag{13.10}$$

and set

$$u^{n+\frac{1}{2}}(x) = w((n+1)\Delta t, x).$$

Step 2.

$$u^{n+1}(x) = \inf\left(u^{n+\frac{1}{2}}(x), \psi((n+1)\Delta t, x)\right).$$

It is quite simple to prove the convergence of a splitting method using the Barles–Souganidis framework. There are many VI arising in quantitative finance, in particular in presence of singular controls, and splitting methods are extremely useful for this type of HJB equations. We refer to the guest lecture by H. M. Soner for an introduction to singular control and its applications.

References

1. P. Bank, D. Baum, Hedging and portfolio optimization in financial markets with a large trader. Math. Finance. **14**, 1–18 (2004)
2. G. Barles, Solutions de Viscosité des Équations de Hamilton-Jacobi. *M*athématiques & Applications (Springer, New York, 1994)
3. G. Barles, E.R. Jakobsen, Error bounds for monotone approximation schemes for parabolic Hamilton-Jacobi-Bellman equations. Math. Comput. **76**, 1861–1893 (2007)
4. G. Barles, P.E. Souganidis, Convergence of approximation schemes for fully nonlinear second order equations. Asymptot. Anal. **4**, 271–283 (1991)
5. G. Barles, C. Daher, M. Romano, Convergence of numerical schemes for parabolic equations arising in finance theory. Math. Model Meth. Appl. Sci. **5**, 125–143 (1995)
6. J.-M. Bismut, Conjugate convex functions in optimal stochastic control. J. Math. Anal. Appl. **44**, 384–404 (1973)
7. F. Bonnans, H. Zidani, Consistency of generalized finite difference schemes for the stochastic HJB equation. Siam J. Numer. Anal. **41**, 1008–1021 (2003)
8. U. Cetin, R. Jarrow, P. Protter, Liquidity risk and arbitrage pricing theory. Finance Stochast. **8**, 311–341 (2004).
9. U. Cetin, M. Soner, N. Touzi, Option hedging under liquidity costs. Finance Stochast. **14**(3), 317–341 (2010).
10. U. Cetin, R. Jarrow, P. Protter, M. Warachka, Pricing options in an extended black-scholes economy with illiquidity: theory and empirical evidence. Rev. Financ. Stud. **19**, 493–529 (2006).
11. P. Cheridito, M. Soner, N. Touzi, Small time path behavior of double stochastic integrals, and application to stochastic control. Ann. Appl. Probab. **15**, 2472–2495 (2005)
12. P. Cheridito, M. Soner, N. Touzi, The multi-dimensional super-replication problem under gamma constraints. Annales de l'Institut Henri Poincaré, Série C: Analyse Non-Linéaire. **22**, 633–666 (2005)
13. P. Cheridito, M. Soner, N. Touzi, N. Victoir, Second order backward stochastic differential equations and fully non-linear parabolic pdes. Comm. Pure Appl. Math. **60**, (2007)
14. M.G. Crandall, H. Ishii, P.-L. Lions, User's guide to viscosity solutions of second order partial differential equations. Bull. Am. Math. Soc. **27**, 1–67 (1992)
15. J. Dugundji, Topology (Allyn and Bacon series in Advanced Mathematics, Allyn and Bacon edt.) (1966)
16. N. ElKaroui, M. Jeanblanc-Picqué, *C*ontrôle de processus de markov, Séminaire de Probabilités Springer Lecture Notes in Math. vol. 1321 (1988), pp. 508–541
17. N. ElKaroui, R. Rouge, Pricing via utility maximization and entropy. Math. Finance. **10**, 259–276 (2000)

N. Touzi, *Optimal Stochastic Control, Stochastic Target Problems, and Backward SDE*, Fields Institute Monographs 29, DOI 10.1007/978-1-4614-4286-8,

18. N. ElKaroui, S. Peng, M.-C. Quenez, Backward stochastic differential equations in fiannce. Math. Finance. **7**, 1–71 (1997)
19. A. Fahim, N. Touzi, X. Warin, A probabilistic numerical method for fully nonlinear parabolic PDEs. **21**(4), 1322–1364 (2011)
20. W.H. Fleming, H.M. Soner, *Controlled Markov Processes and Viscosity Solutions* (Springer, New York, 1993)
21. H. Foellmer, P. Leukert, Quantile hedging. Finance Stochas. **3**, 251–273 (1999)
22. Y. Hu, P. Imkeller, M. Müller, Utility maximization in incomplete markets. Ann. Appl. Probab. **15**, 1691–1712 (2005)
23. I. Karatzas, S. Shreve, *Brownian Motion and Stochastic Calculus*, 2nd edn. (Springer, New York, 1991)
24. I. Karatzas, S. Shreve, *Methods of Mathematical Finance*, 2nd edn. (Springer, New York, 1999)
25. N. Kazamaki, *Optimal Stopping and Free-Boundary Problems*. Lectures in Mathematics (Springer, Berlin 1579)
26. M. Kobylanski, Backward stochastic differential equations and partial differential equations with quadratic growth. Ann. Appl. Probab. **28**, 558–602 (2000)
27. N.V. Krylov, *Controlled Diffusion Processes* (Springer, New York, 1977)
28. H. J. Kushner, P.G. Dupuis, *Numerical Methods for Stochastic Control Problems in Continuous Time* (Springer, New York, 1992)
29. R.C. Merton, Lifetime portfolio selection under uncertainty: the continuous-time model. Rev. Econ. Stat. **51**, 247–257 (1969)
30. R.C. Merton, Optimum consumption and portfolio rules in a continuous-time model. J. Econ. Theor. **3**, 373–413 (1971)
31. M.-A. Morlais, *Equations Différentielles Stochastiques Rétrogrades à Croissance Quadratique et Applications*. Ph.D. thesis, Université de Rennes, 2007
32. E. Pardoux, S. Peng, Adapted solution of a backward stochastic differential equation. Syst. Contr. Lett. **14**, 55–61 (1990)
33. D. Pooley, P. Forsyth, K. Vetzal, Numerical convergence properties of option pricing pdes with uncertain volatility. IMA J. Numer. Anal. **23**, 241–267 (2003)
34. P.J. Reny, On the existence of pure and mixed strategy nash equilibria in discontinuous games. Econometrica. **67**, 1029–1056 (1999)
35. R.T. Rockafellar, *Convex Analysis* (Princeton University Press, Princeton, 1970)
36. H.M. Soner, N. Touzi, Dynamic programming for stochastic target problems and geometric flows. J. Eur. Math. Soc. **4**, 201–236 (2002)
37. M. Soner, N. Touzi, The dynamic programming equation for second order stochastic target problems. SIAM J. Contr. Optim. **48**(4), 2344–2365 (2009)
38. H.M. Soner, N. Touzi, J. Zhang, Wellposedness of second order backward sdes. Probability Theory and Related Fields **153**, 149–190 (2012)
39. R. Tevzadze, Solvability of backward stochastic differential equation with quadratic growth. Stochast. Process. Appl. **118**, 503–515 (2008)
40. J. Thomas, *Numerical Partial Differential Equations: Finite Difference methods* (Texts in Applied Mathematics) (Springer, New York, 1998)
41. J. Wang, P. Forsyth, Maximal use of central differencing for hamilton-jacobi-bellman pdes in finance. Siam J. Numer. Anal. **46**, 1580–1601 (2008)
42. T. Zariphopoulou, A solution approach to valuation with unhedgeable risks. Finance Stochast. **5**, 61–82 (2001)
43. R. Zvan, K.R. Vetzal, P.A. Forsyth PDE methods for pricing barrier options. J. Econom. Dynam. Control **24**, 1563–1590 (2000)

MIX
Papier aus verantwortungsvollen Quellen
Paper from responsible sources
FSC® C105338

If you have any concerns about our products,
you can contact us on
ProductSafety@springernature.com

In case Publisher is established outside the EU,
the EU authorized representative is:
Springer Nature Customer Service Center GmbH
Europaplatz 3, 69115 Heidelberg, Germany

Printed by Libri Plureos GmbH
in Hamburg, Germany